3800 14 0052815 8
D1134419

READER'S DIGEST DIY PLUMBING & HEATING

READER'S DIGEST
DIY
PLUMBING & HEATING

STEP BY STEP INSTRUCTIONS • EXPERT GUIDANCE • HELPFUL TIPS

Published by
The Reader's Digest Association, Inc.
London • New York • Sydney • Montreal

Contents

Introduction

Understanding the system

Maintenance and repairs

Plumbing emergencies – what to do if:

Water pours from the loft

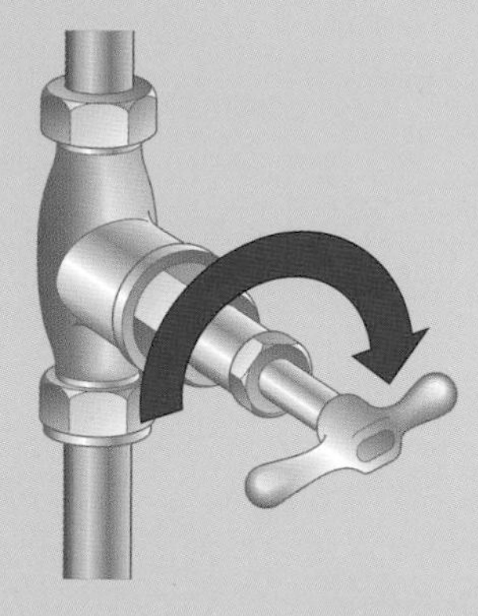

1 Turn the main stoptap off (clockwise). It is usually close to the kitchen sink (if you can't, see right). Put buckets under the leaks, then turn on all the cold taps in the house and flush all the WCs to drain the cold water storage cistern.

2 Find the cause of the trouble. It may be a burst pipe in the loft or a cistern overflow caused by a blocked overflow pipe.

No water comes from a tap

1 If no water flows from the kitchen sink cold tap, check that the main stoptap is open. If it is, call your water supply company. You will find their emergency number on your water bill.

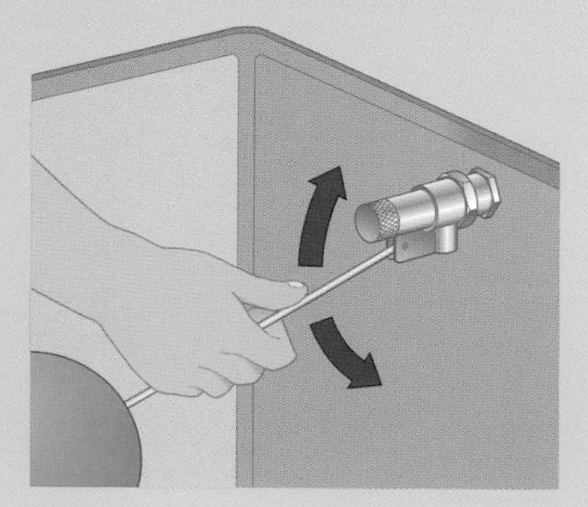

2 If no water flows from other taps, check the cold water cistern. It may have emptied because of a jammed or blocked ballvalve. If it is empty, move the float arm sharply up and down to free the valve, then clean the valve (page 48).

Alternatively In frosty weather there may be an ice plug blocking a supply pipe. If the kitchen cold tap is working, check the flow into the cold water cistern by pressing down the ballvalve. If there is no inflow, the rising main is frozen, probably between the ceiling and the cistern inlet.

3 If the cistern is filling, check the bathroom taps. If there is no flow from one tap, its supply pipe from the cistern is frozen.

4 To thaw a pipe, strip off any lagging from the affected part and apply hot water bottles. If a pipe is difficult to get at, blow warm air onto it with a hair dryer.

WARNING Do not use a blowtorch to defrost a frozen pipe. It may cause a fire, or melt the solder in a pipe joint and cause another leak.

Hot water cylinder leaks

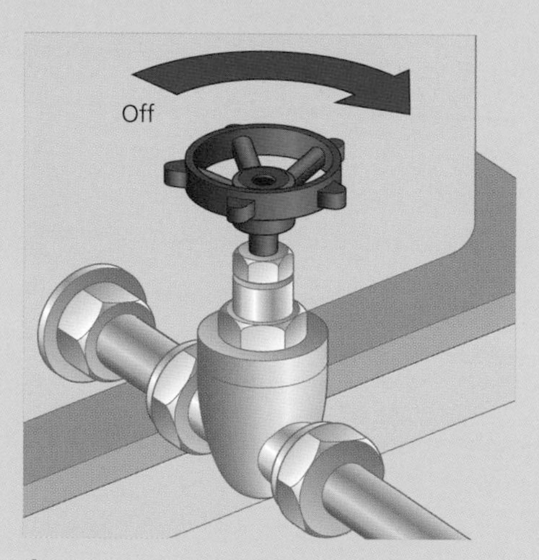

1 Turn off the gatevalve (clockwise) on the supply pipe from the cold water cistern to the hot water cylinder. If there is no gatevalve, turn off the main stoptap and turn on all the taps to empty the cistern. (This will not empty the hot water cylinder, but will stop water from flowing into it.)

2 Switch off immersion heater, if fitted.

3 Switch off the boiler, or put out the boiler fire.

4 Connect a hose to the cylinder drain valve, which is located near the base of the cylinder where the supply pipe from the cold water cistern enters. Put the other end of the hose into an outside drain.

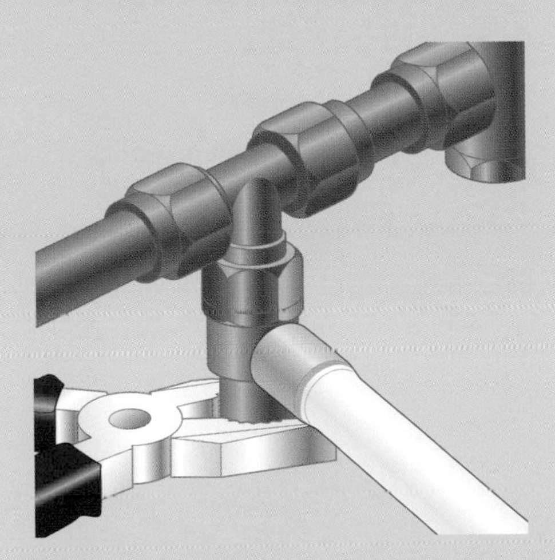

5 Open up the drain valve with a drain valve key or pliers.

6 Get the hot water cylinder repaired or replaced by a plumber.

You cannot turn off the water

If the problem is on the cistern-fed hot or cold supplies, you can tie up the float valve in the cistern, then run the taps and flush the WCs to empty the pipes. If the problem is on the mains fed cold supply, or hot supply through an unvented cylinder or combination boiler, you need to turn off the rising main either at the stoptap inside your home or the outdoor stoptap in the ground (page 12). You will probably need a key to do this. If you don't have one, call your water company for assistance.

Other plumbing problems

Electrical work

A small proportion of plumbing work in the home involves working with mains electricity. The two obvious examples are installing an electric shower (or a shower with a pump) and the wiring involved in a central heating installation.

New – or replacement – electric wiring in the home is now included in the Building Regulations (as Part P – see page 9) and has to meet the conditions of the 17th Edition of the Wiring Regulations (BS7671).

All of this is explained in detail in our companion book *Wiring & Lighting*, but what you need to know is that any wiring work you carry out which involves a new circuit, or which is in a kitchen or bathroom, needs to be notified to the Building Control Department of your local authority (see page 9 for details).

Part-P registered electricians do not need to notify the local authority. They can "self-certify" their own work – which must, of course, be carried out following the latest Wiring Regulations.

Details of the two main electricians' trade associations are given on page 121.

Gas safety

Mains gas will not poison you, but it can explode if it leaks and is ignited. It can also kill indirectly if it is not burned safely and under controlled conditions in a gas fire, boiler or water heater.

Follow these guidelines to ensure that you always use gas safely.

Gas safety regulations

The Gas Safety (Installation and Use) Regulations make it illegal for anyone to carry out work relating to gas supply and fittings who is not 'competent'. In practice, this means leaving all work on your gas supply system to a qualified gas fitter who is on the Gas Safe Register (www.gassaferegister.co.uk) covering Great Britain and the Isle of Man.

1 Never attempt DIY work on your gas pipes, fittings or appliances: this is illegal. Always call in a Gas Safe registered fitter to do the work.

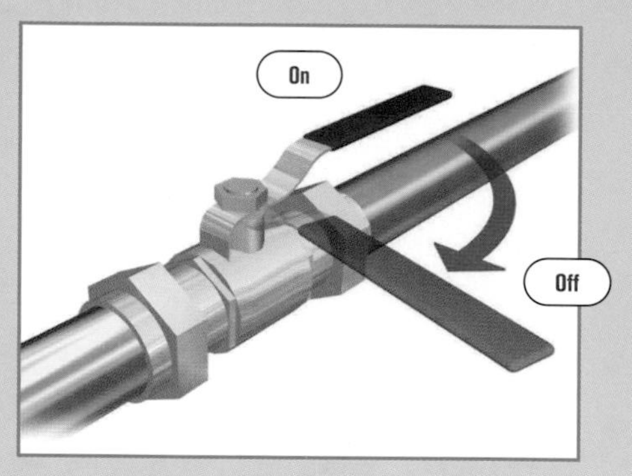

2 If you smell gas turn off the supply at the main on/off lever (you'll find it next to the gas meter) by moving it so it is at right angles to the pipe. Open all doors and windows. Put out all naked lights and extinguish cigarettes with water. Do not turn any electrical switches either on or off as this can create a spark. Contact the National Grid Emergency Service immediately (0800 111 999) if the smell of gas persists, and call a registered gas fitter to trace the fault if it disperses.

3 Buy only gas appliances that comply with British or European standards.

4 Have gas appliances serviced at least annually by a qualified Gas Safe registered fitter. Look out for danger signs. If there is sooting round an appliance, if it burns with an orange or lazy flame, or if there is excessive condensation nearby, it may be faulty. Call a Gas Safe registered fitter as soon as possible and stop using the appliance.

5 Ensure that rooms containing gas-burning appliances are properly ventilated. If you get headaches or nausea when they are operating, they may be burning fuel unsafely and creating potentially lethal carbon monoxide (CO) gas. Ask your gas supplier for advice if you are concerned.

6 For complete safety, have a carbon monoxide detector installed – ideally a mains-powered one made to British Standard **BS EN 50291-1:2010** – look for the British Standards Kitemark (right). This is particularly important if you have gas heaters in bedrooms.

Rules for plumbing and heating

When making changes or additions to your home's plumbing or heating, there are two sets of rules that you must adhere to. They are the Water Regulations and the Building Regulations. If you are in any doubt as to whether these regulations apply to you, contact the authority concerned.

Water Regulations

Plumbing in new houses is covered by the Water Supply (Water Fittings) Regulations which are enforced by local water companies. The regulations, which also apply to extensions and alterations to existing plumbing systems, are designed to prevent the waste, undue consumption, misuse or contamination of water supplies – and you can be fined if you contravene them.

Provided you follow the regulations, a copy of which can be downloaded from the Government's legislation website (www.legislation.gov.uk), you do not need to tell anyone what you are doing. But you do have to inform your water supplier if you want to install a water softener (see page 21) or a bidet with an ascending spray (this does not apply to a normal bidet, see page 68). You must wait for their consent before starting work.

There is an illustrated Guide to the Regulations available from the Water Regulations Advisory Scheme – www.wras.co.uk.

Building Regulations

The Building Regulations apply to both new houses and any extensions and alterations to existing houses. You can download a copy of the regulations (and the very useful explanatory Approved Documents for each part) at the Government's Planning Portal website (www.planningportal.gov.uk).

The Building Regulations for your area are administered by the Building Control Department of your local authority and you must contact them if you are doing any work that falls within the regulations. A fee is payable if you need to secure Building Regulations approval.

The main regulations that affect the home plumber are:

- Part H (Drainage and waste disposal) – all soil, waste and underground drainage
- Part J (Combustion appliances and fuel storage) – mainly affecting the ventilation required for new boilers
- Part L (Conservation of fuel and power) – mainly affecting central heating controls and new boilers
- Part P (Electrical Safety) – mainly affecting new electric shower circuits, new central heating control wiring installations and anything to be plumbed into a kitchen or bathroom, which involves extending wiring.

Understanding the system

How water is supplied to the home

Whether for home improvements, or for tackling emergencies, it is important to know what type of water system you have, and where to find all the relevant system controls.

The cold water supply

There are two types of cold water supply in British homes: direct and indirect.

In a direct cold water supply, branch pipes from the rising main lead directly to all the cold taps and WC cisterns in the house. This means that you can drink cold water from any tap. A pipe from the rising main may feed a storage cistern in the loft – to feed the hot water cylinder – or the cylinder may have its own mini-cistern on top. More often, hot water comes from a combination boiler (page 14).

Most British homes have an indirect system. The rising main feeds the cold tap at the kitchen sink (and possibly pipes to a washing machine and an outside tap). This water is clean drinking water. It then continues up to a cold water storage cistern in the roof, which supplies all other taps, the WCs and the hot water cylinder.

There are advantages to an indirect system: water from a cold water storage cistern gives even water pressure, which produces quieter plumbing and less wear and tear on washers and valves. Leaks are also less likely, and any leak that does occur will be less damaging than one from a pipe under mains pressure.

Water from a cistern is warmer than mains water, so less hot water is needed for washing and bathing. It also reduces condensation on WC cisterns. And if the house supply is temporarily cut off – for work on the mains, for example – there is a supply of stored water available for use.

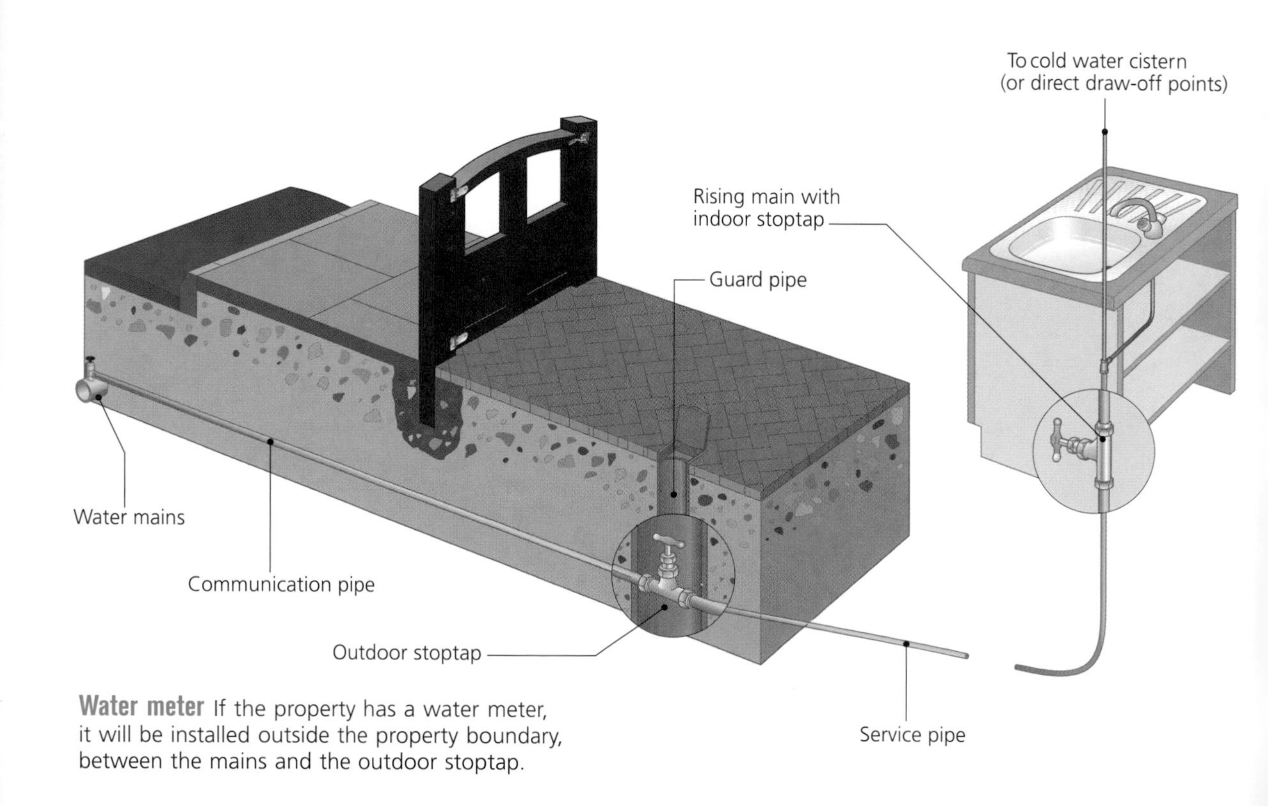

Water meter If the property has a water meter, it will be installed outside the property boundary, between the mains and the outdoor stoptap.

Water mains The water supply to most British homes is provided by the local water supply company, through iron or heavy plastic water mains.

Communication pipe From the mains, a pipe known as a communication pipe takes the water to the water company's outdoor stoptap – a control valve about 1m below the ground at or near the boundary of each property.

Outdoor stoptap The stoptap, which is turned with a long key, is at the bottom of an earthenware guard pipe under a small metal cover, set into the surface of the garden or the public footpath outside. In older properties, this may be the only place where the water can be turned off.

Service pipe From the water company's stoptap, a service pipe carries water into the house. The pipe should meander slightly in the trench to allow for ground movement, which would otherwise pull on the fittings at each end. To avoid frost damage, it should be at least 750mm and not more than 1.35m below ground.

Rising main The service pipe enters the house, usually close to the kitchen sink (but sometimes under the stairs or in a garage). From there, the rising main leads to the cold water storage cistern in the loft and/or direct draw-off points. New houses have a 25mm MDPE (plastic) service pipe to allow for the use of unvented cylinders or combination boilers. If you have a lead service pipe contact your water supply company to ask about their lead replacement programme.

SAFETY NOTE

If your house has an indirect system, do not drink water from any tap other than the kitchen one. Water from a cistern may not be clean.

BE PREPARED

Make sure that you and others in the house know where the indoor and outdoor stoptaps are, as well as the gatevalves on the supply pipes to the hot water cylinder and cold taps, and label them.

FAULT DIAGNOSIS

LEAKING PIPE

Burst or leaking pipe **Turn off water and drain cold water cistern (see page 29). Then repair pipe (see page 32).**

NO WATER FROM TAPS

No supply from rising main **Check main stoptap is open. If it is, contact water company. You can find the number under 'Water' in the phone book.**

Jammed ballvalve in cistern **Check whether cistern is filling. If not, depress the float arm to free the valve and fill the cistern. If this fails, replace the ballvalve. See page 52.**

Check for air locks **To clear an airlock, see page 34.**

OVERFLOW DRIPPING

Faulty ballvalve **Service or replace ballvalve. See pages 48-52.**

Float Arm too high **Adjust or bend arm downwards to lower water level in cistern. See page 48.**

FROZEN PIPES

No lagging or temperature dropped too low **Defrost pipes and insulate. See page 100.**

BUILD-UP OF LIMESCALE

You live in hard water area *Install a water treatment unit (see page 20).*

Hot water supply

There are two main ways of providing hot water in the modern home: storage systems and instantaneous systems. Most homes will have one system or the other; some homes may have both. What you have will depend on the size and age of your home and the type of cold water supply you have.

Homes with an indirect cold water supply (where cold water is supplied from a large storage cistern in the loft) are most likely to have a storage hot water system, with the hot water coming from a hot water cylinder.

Homes with a direct cold water supply (all cold taps fed directly from the rising main) are more likely to have an instantaneous hot water system with hot water being supplied either by a gas multipoint water heater or by a combination boiler.

Few homes these days are likely to have back boilers or boilers which heat only the hot water and not a circuit of radiators. These 'direct' boiler systems suffered badly from scale and should be replaced where possible.

Storage hot water systems

The central feature of a hot water storage system is the hot water cylinder, normally fed by cold water at its bottom by a pipe leading from the cold water storage cistern and with a pipe (the vent pipe) coming out of its top which leads back up to the cold water storage cistern, curving up and over the top of it with the end passing through the lid and finishing just above the water level. A branch pipe just above the hot water cylinder takes hot water to all the house hot taps: the vent pipe is there as a safety measure to allow an escape route for air or if the water in the cylinder overheats.

The water in the cylinder can be heated in one or both of two ways: immersion heater or heating coil.

Immersion heater An immersion heater is a bit like the element in a kettle. It has its own electric supply circuit (with an on/off switch close to the cylinder) and when it is turned on it heats the water in the cylinder. An immersion heater may be the only method of heating the hot water (in an all-electric house or one with warm-air heating) as shown opposite or it may be in addition to a heating coil as shown on page 16. See page 95 for more details.

Heating coil This is a coil of pipework fitted within the hot water cylinder through which water heated by the central heating boiler passes – see drawing on page 16. Heat is transferred through the walls of the pipe coil to the water in the cylinder, but the water from the boiler (in the 'primary' circuit) does not mix with the water coming from the cold water cistern and going (after it's heated) to the hot taps – known as the 'secondary' circuit.

Unvented hot water systems These are the latest systems available for domestic hot water heating. What makes them different from normal systems is that the cold water supply to the cylinder comes directly from the rising main – so if all the cold taps in the house are also fed from the rising main, there is no need for a cold water storage cistern in the loft.

Special cylinders are needed (because of the higher pressure) plus a whole host of safety devices (which makes their installation a job for a plumber), but they can have either immersion heaters or heating coils just like normal ('vented') systems.

If the central heating boiler uses a 'sealed' system (see page 19), there is no need for a header tank in the loft either.

Instantaneous hot water systems

A whole-house instantaneous hot water system will use either a gas multi-point water heater or a combination ('combi') boiler. Both take water directly from the rising main and heat it as it is needed, distributing it around the house to the hot water taps. The combination boiler (see illustration, right) also heats the radiators.

In a large house, you might use a gas multi-point in addition to a storage system for providing hot water to rooms that are a long way from the hot water cylinder.

There are also point-of-use instantaneous water heaters for heating water where it is needed – gas or electric water heaters over a kitchen sink, for example – or powerful electric heaters for showers.

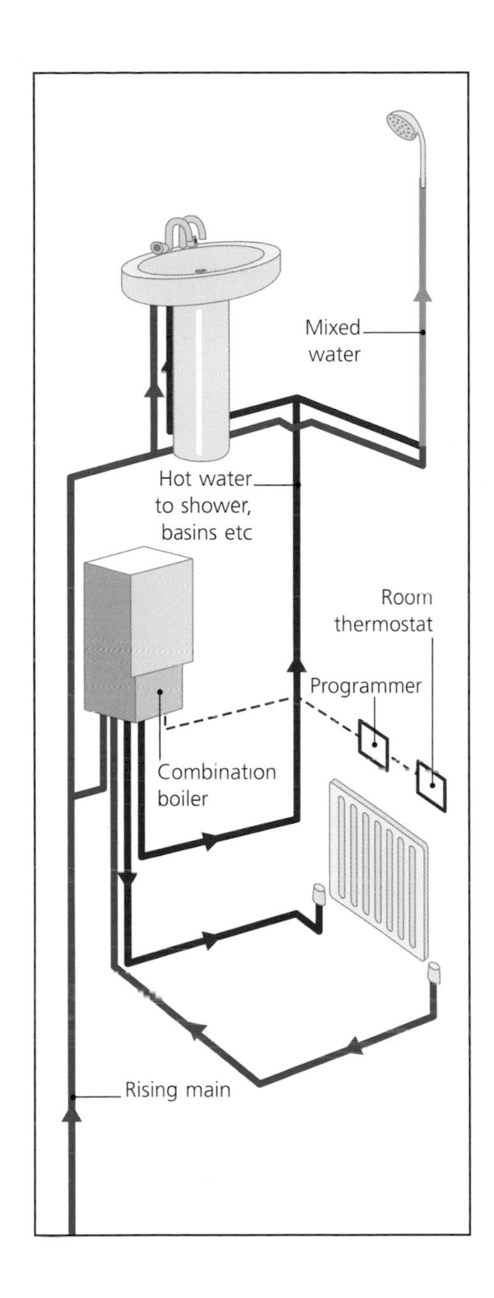

Instantaneous hot water The combination boiler shown here heats cold water directly from the rising main and also supplies heated water to radiators.

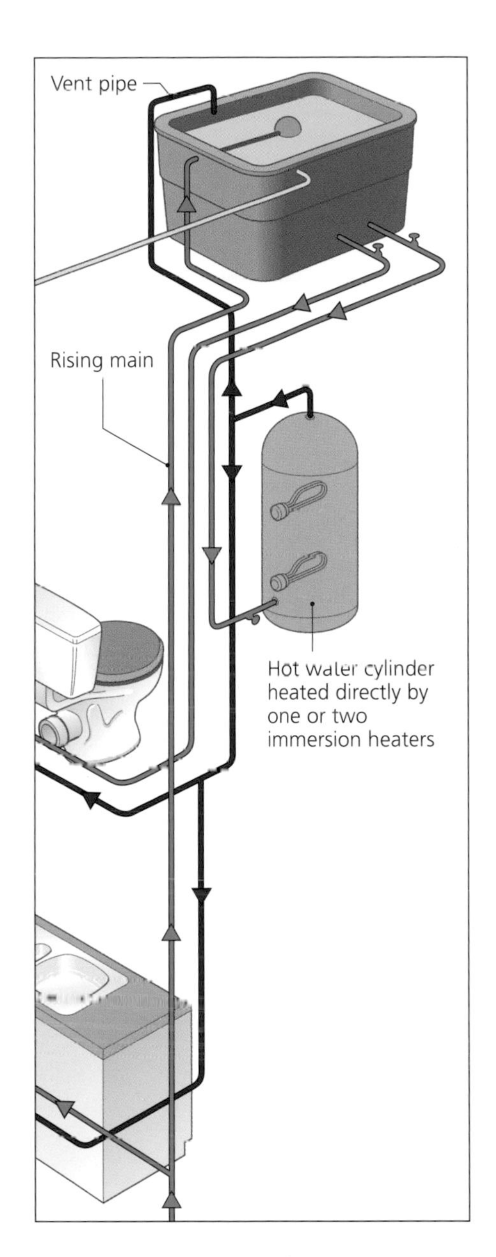

Storage hot water The hot water cylinder here is fitted with immersion heaters to heat water up from cold, but is not connected to a central heating boiler.

A typical household plumbing system

1 The rising main feeds the kitchen cold tap and kitchen appliances before it rises to the cold water storage cistern and the central heating header tank. It also feeds an electric (instantaneous) shower.

2 The cold water cistern is filled from the rising main, and the inflow of water is controlled by a float-operated ballvalve. The capacity of the average household cistern is 230 litres, and it has an overflow pipe to carry water out to the eaves if the cistern overfills through a failure of the ballvalve.

3 Water regulations require new cold water storage cisterns and header tanks (feed-and-expansion cisterns) to have dust-proof and insect-proof (but not airtight) covers and to be insulated against frost. The main cold water storage cistern must supply drinkable water.

4 Water from the cistern is distributed by at least two 22mm (or 28mm) diameter pipes fitted about 75mm from the bottom. One supplies WCs and cold taps – except for the kitchen cold tap. Normally the 22mm pipe goes direct to the bath cold tap, and 15mm branches feed washbasins and WC cisterns.

5 The other distribution pipe feeds cold water to the bottom of the hot water cylinder – usually a copper cylinder of about 140 litres capacity. In a typical modern house with a central heating boiler, there are two water circuits through the hot water cylinder – the primary circuit that heats the water, and the secondary circuit that distributes it.

6 Hot water stored in the cylinder is heated by the primary water circuit through the boiler, and sometimes also by an immersion heater. The boiler can also heat radiators.

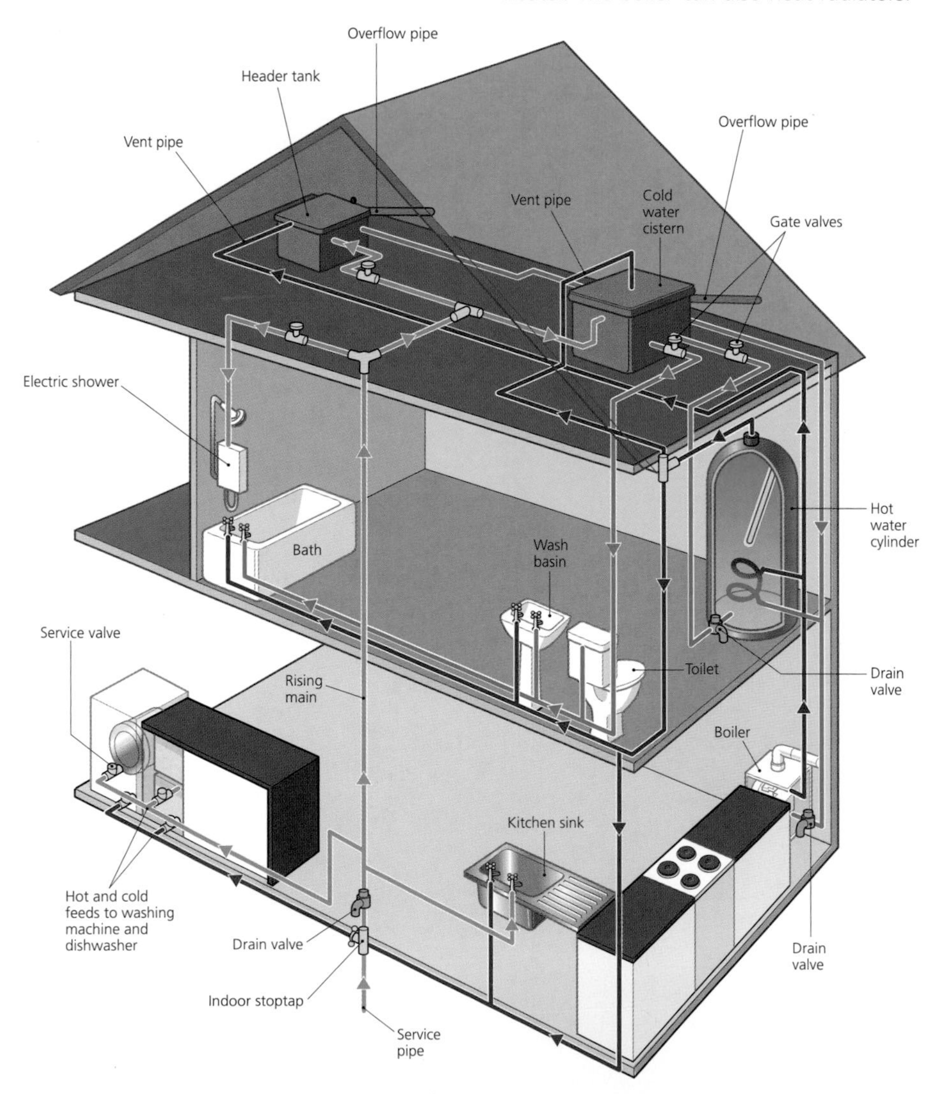

Where the waste water goes

If you live in a house built before the mid-1960s, you probably have a two pipe drainage system; newer houses have one drain pipe – a single stack system.

Whatever the drainage system, every bath, basin or sink in the house is fitted with a trap – a bend in the outlet pipe below the plughole. This holds sufficient water to stop gases from the drains entering the house and causing an unpleasant smell. The trap has some means of access for clearing blockages. All WC pans have built-in traps.

Below ground, the household waste pipes or drains are channelled through an inspection chamber near the house to form the main drain, which runs into the water company's sewer.

Single stack system

Modern houses have a single stack drainage system. Waste from all sinks and WCs is carried underground by a single vertical pipe known as a soil stack. This pipe may be installed inside the house and its vented top extends above the roof.

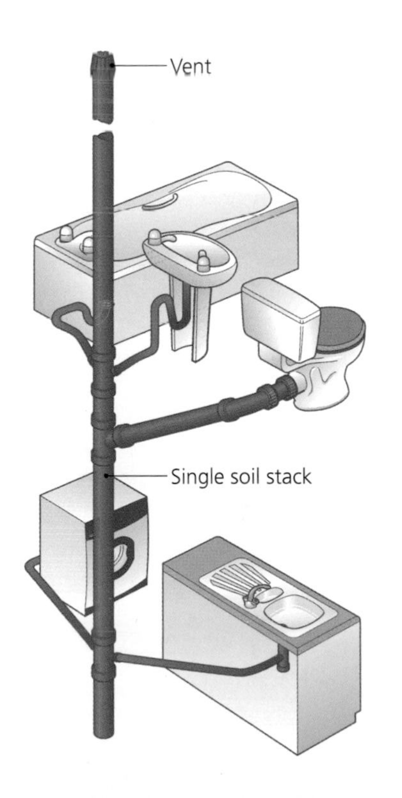

Two pipe system

Most houses built before the mid-1960s have what is known as a two pipe drainage system for waste water disposal.

A vertical soil pipe fixed to an outside wall carries waste from upstairs WCs to an underground drain.

The open top of the soil pipe – the vent – extends above the eaves and allows the escape of sewer gases. It is protected from birds with a plastic terminal guard.

Ground floor WCs usually have an outlet direct into the underground drain.

A second outside pipe – the waste pipe – takes used water from upstairs baths, basins and showers via an open hopper head to empty into a ground-level gully. Water from the kitchen sink also runs into a gully.

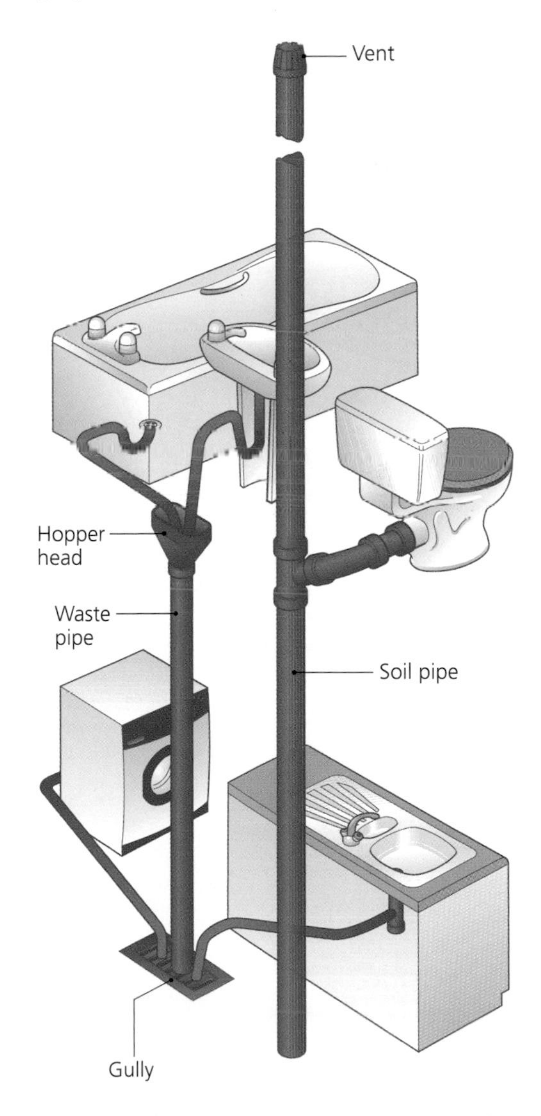

How central heating works

Most central heating systems warm the rooms of a house by passing hot water through a system of pipes and radiators. The type of central heating system you have may depend on the age of your house. It is important to understand how it works.

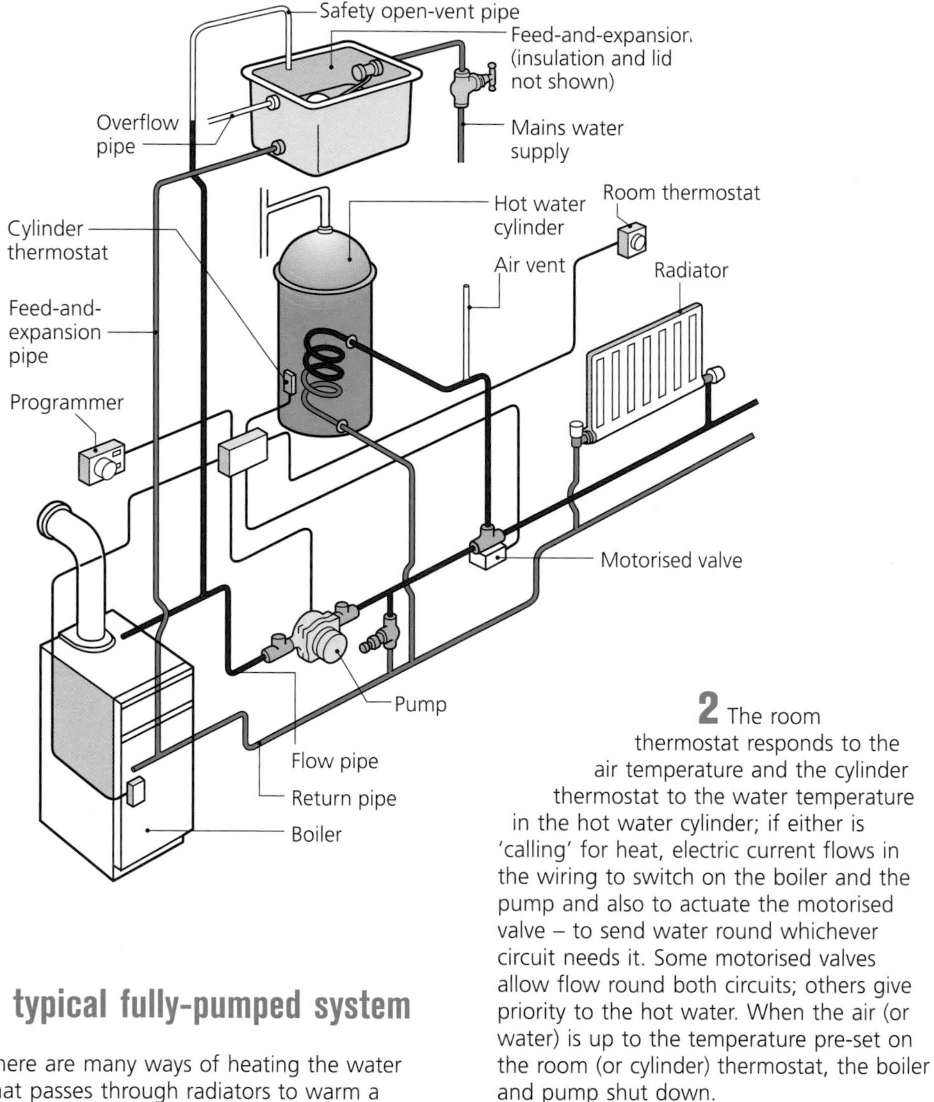

A typical fully-pumped system

There are many ways of heating the water that passes through radiators to warm a house, but it is usually heated by a boiler, which switches on automatically at certain times of day.

1 The 'central' in central heating is the boiler – but it is the programmer that is in charge. This has two channels – one for heating and one for hot water – which provide the electrical energy at pre-set times of day to switch the boiler and the pump on to send heated water round the radiator and hot water circuits.

2 The room thermostat responds to the air temperature and the cylinder thermostat to the water temperature in the hot water cylinder; if either is 'calling' for heat, electric current flows in the wiring to switch on the boiler and the pump and also to actuate the motorised valve – to send water round whichever circuit needs it. Some motorised valves allow flow round both circuits; others give priority to the hot water. When the air (or water) is up to the temperature pre-set on the room (or cylinder) thermostat, the boiler and pump shut down.

3 The same water is constantly circulated around the system. In an open vented system, in case of leakage or evaporation, the water is topped up from a feed-and-expansion cistern (header tank). This cistern also takes up the expansion that occurs when the water heats up from cold.

4 An open-ended pipe, called the safety open-vent pipe, provides an escape route for steam and excess pressure if the boiler overheats.

Gravity hot water circulation

In some older central heating systems and in solid fuel systems, the heated hot water going to the hot water cylinder is circulated by gravity. When water is heated it expands and hot water weighs less than cold water.

1 Hot water rises up a large pipe from the boiler to the hot water cylinder. Cooled water descends down the return pipe, pushing the lighter hot water up the flow pipe. Gravity circulation is reliable as it needs no mechanical assistance, but it requires larger 28mm pipes. The system is most efficient if the cylinder is directly above the boiler.

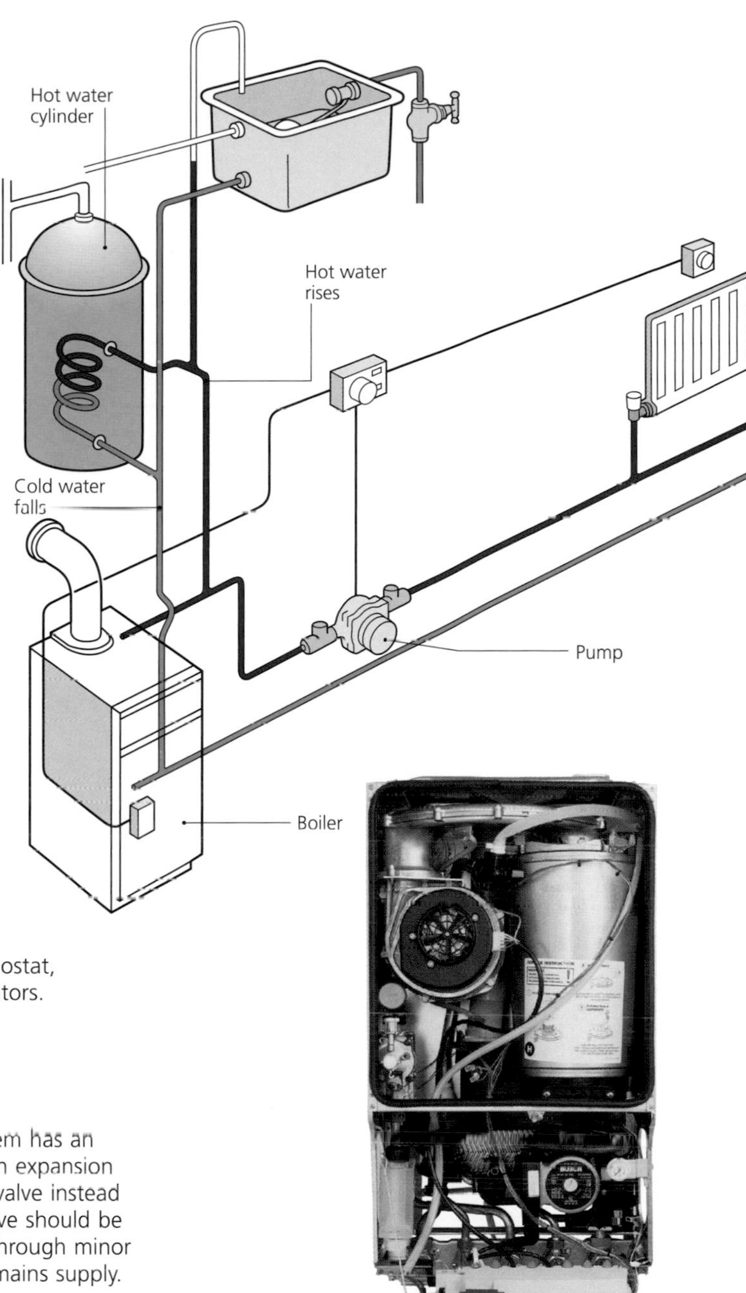

2 A pump, controlled by a programmer and room thermostat, drives water around the radiators.

Combination boilers are a compact and energy-saving option. All the essential components are housed inside.

A sealed system

A sealed central heating system has an expansion vessel instead of an expansion cistern, and a pressure relief valve instead of a safety-vent pipe. The valve should be set to 3 bar. Any water lost through minor leaks is topped up from the mains supply.

1 The system is controlled by a programmer as for a full-pumped system.

2 A room thermostat starts the boiler and pump to send water round the radiator circuit and a cylinder thermostat does the same for the hot water circuit (except for combination boilers – see below).

3 The boiler has an over-heat cut-out to prevent the system boiling should the standard thermostat fail, and on no account must a boiler without over-heat protection be fitted to a sealed system. A 'system' boiler used in sealed systems has many of the necessary components housed within the boiler casing.

Combination boilers All combination boilers utilise sealed systems. As well as saving space because of the lack of a feed-and-expansion cistern, a combination boiler also has the advantage that no hot water cylinder is needed as the boiler heats mains water and delivers hot water directly to the taps at mains pressure.

Softening your water

What makes water 'hard' is the dissolved calcium and magnesium salts it can contain. When the water is heated, these 'precipitate' out and form scale – especially on the heating elements of kettles, washing machines and electric showers. A water softener replaces these salts with sodium salts that do not form scale.

Water softeners use an ion-exchange system (see below) to replace the scale-forming salts with harmless sodium ones. An alternative (for showers and combi boilers) is a plumbed-in scale inhibitor – see page 94 for more details.

The most common plumbed-in system is attached to the incoming water supply so that water used for bathing and washing clothes is softened, but drinking water remains untreated (this is especially important for babies and people on a sodium-restricted diet). This type of system works by passing the water ❶ through resin beads ❷ that replace the scale-forming magnesium ❸ and calcium ❹ ions with harmless and taste-free sodium ions ❺. The beads are regenerated by periodically flushing through with salt water, to replenish the supply of sodium ions and remove the magnesium and calcium ions.

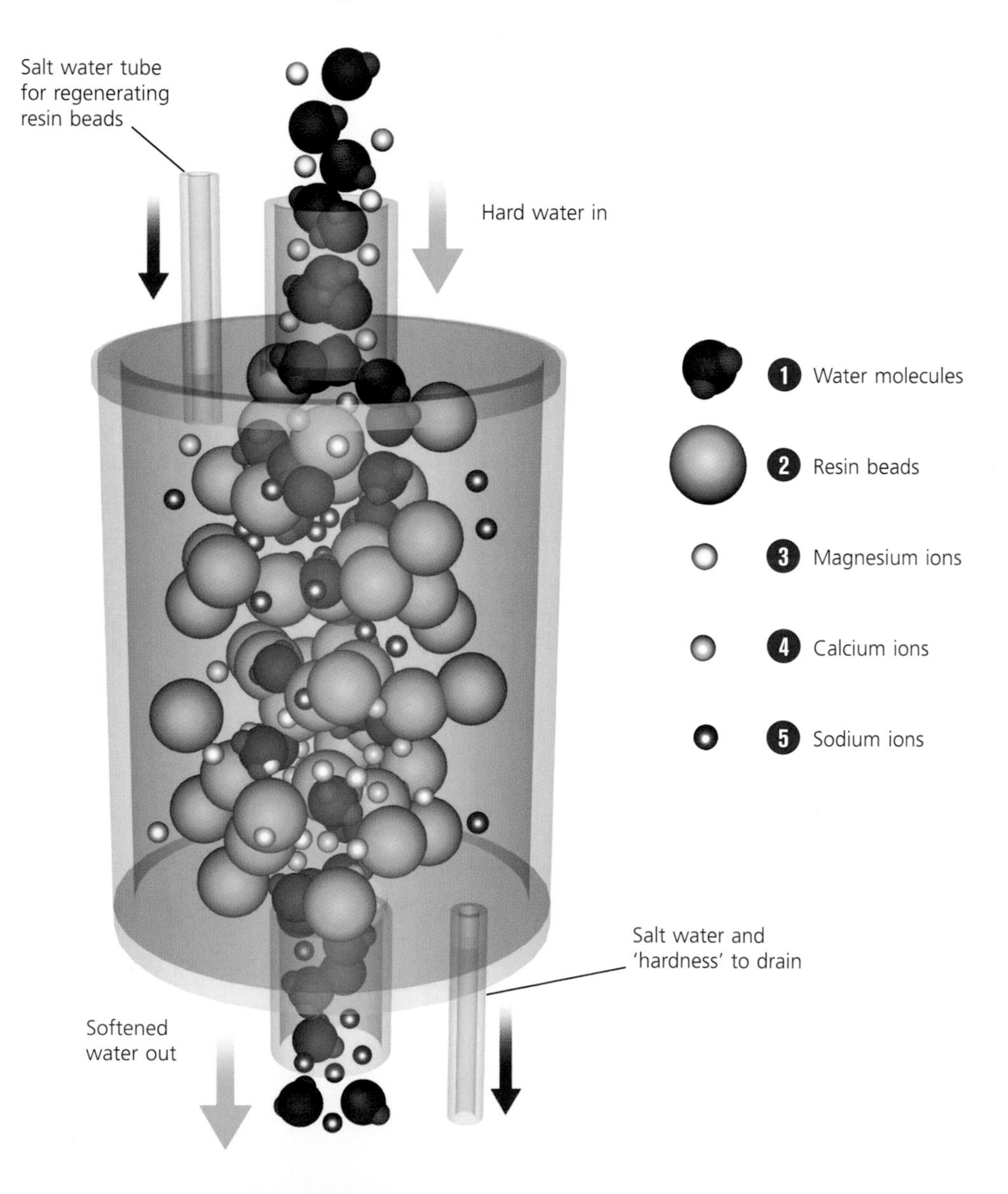

Positioning a water softening unit

Any permanently-fitted device that removes hardness from mains water must be fitted close to the rising main and a drain. Some units contain an electrical system and these require a power source.

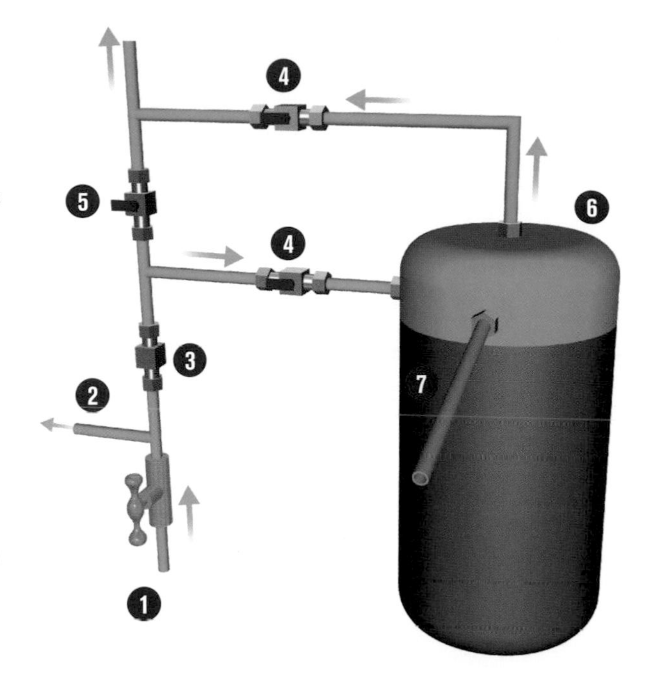

Immediately after the drain valve (not shown) above the main stoptap ❶, a pipe ❷ goes to the kitchen cold tap for drinking (and, preferably, any garden taps) and there must be a non-return valve ❸ to prevent softened water being drawn back through when the drinking water tap is turned on (and to prevent backflow into the mains supply). Servicing valves ❹ are normally open, but can be closed for maintenance; the bypass valve ❺ is usually closed, but is open when the servicing valves are shut. The water softener ❻ has an overflow pipe ❼ and a waste pipe to the drains.

You must inform your water company if you intend fitting a water softener. You will also need to contact Building Control if an electrical connection requires additional wiring in a kitchen or bathroom.

PLUMBING MAINTENANCE

ONCE A MONTH

Turn rising main stoptap on and off several times to prevent seizing
Spray small amount of aerosol lubricant onto spindle. Open tap fully and close by a quarter turn.

BEFORE YOU GO ON HOLIDAY

Turn off the water at indoor stoptap on rising main, and drain cistern **(see page 29)**
This means that there is no stored water in the system to cause damage if pipe bursts.
Leave the heating on low if you go away in winter

BEFORE WINTER

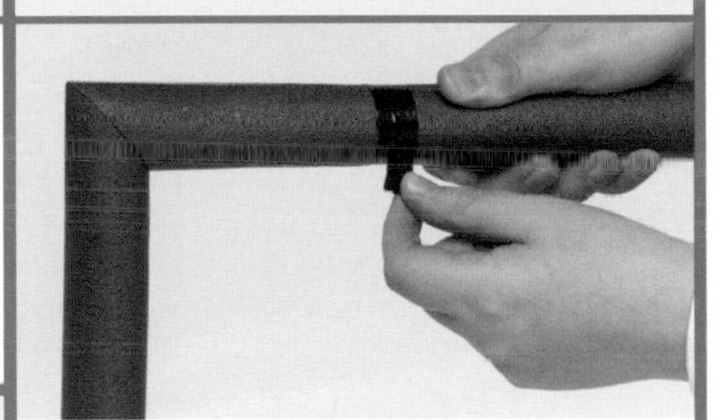

Lag all cold pipes with self-adhesive foam wrap or tube **(see page 100)**
Remember not to insulate the loft floor under cold water cistern. This allows heat from the house to reach the cistern and prevents it freezing in cold weather.
Drain any outside taps *Turn off isolating stoptap and drain the pipe (or leave the tap open).*

Common plumbing and

The plumbing system

Your plumbing system can suffer from a wide range of faults, but they fall into four main categories.

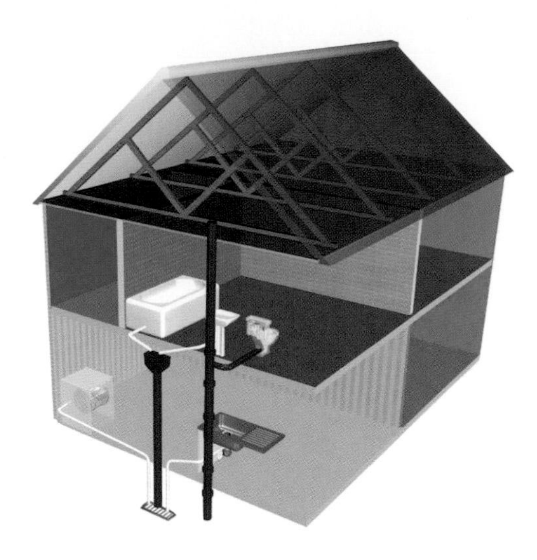

Taps and valves ❶

Taps and valves control the flow of water into and through the home. They may drip, causing stains and overflows, or may jam open or closed, or be hard to operate. Taps might need maintenance or replacement, depending on the severity of the problem. See pages 36-42.

Water pipes ❷

Cold pipes take water to wherever it is needed. They may leak, due to perforations developing in the pipe or faulty seals at pipe connectors. They may also become blocked by scale, caused by hard water. Both faults are easily remedied, although leaks demand swift damage limitation – *see* Plumbing emergency action (right).

Storage cisterns and cylinders ❸

Storage cisterns and cylinders hold cold and hot water respectively. They may develop leaks or, in the case of the hot water cylinder, may become inefficient due to a build-up of limescale.Replacement is the only long-term solution.

Waste pipes

Waste pipes convey used water from appliances and WCs to the household drains, via U-shaped traps designed to keep drain smells out of the house. They can become blocked, causing overflows, but most blockages can be cleared easily. See page 34.

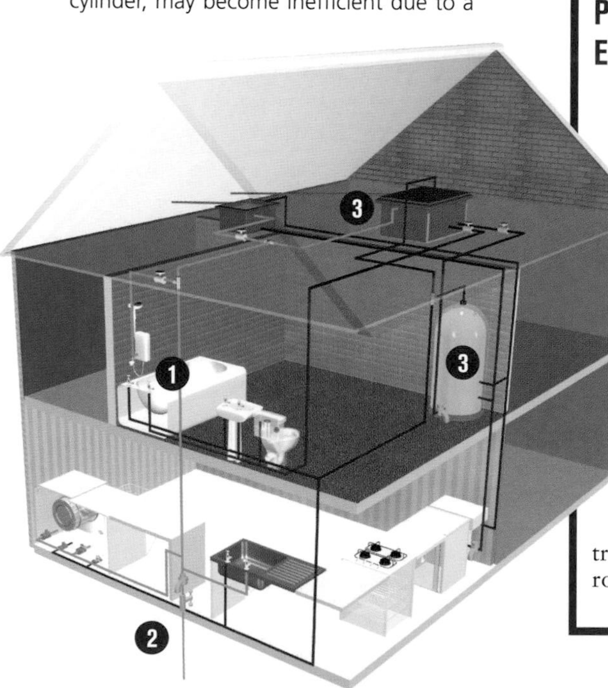

PLUMBING EMERGENCY ACTION

- If a fault causes water to escape, aim to stop the flow as quickly as possible.
- Make sure you know in advance where the system's main on/off and flow controls are located (see page 12).
- Empty leaking water storage cisterns and cylinders and supply pipes by turning off the main stoptap and opening cold taps (see page 29), or by attaching a garden hose to a drain valve and opening the valve with pliers or a spanner. Lead the hose outside the house.
- Clear blocked waste pipes and traps by dismantling, plunging or rodding (see page 34).

heating faults

The heating system

A central heating system consists of five main components, each of which can malfunction or fail.

The boiler ❶
The boiler heats the system (and may also provide domestic hot water). It contains a number of parts that will need regular maintenance and eventual replacement and this is a job for a professional (see page 7). Annual servicing will keep the boiler in good order. A noisy boiler can indicate corrosion – see page 106

The pump ❷
The pump circulates heated water round the system. It may become jammed or noisy in operation, or may simply fail altogether. Regular operation and cleaning help to prevent problems, but replacement is usually straightforward. See page 114.

The radiators ❸
The radiators transmit heat to individual rooms. They sometimes develop pinhole leaks due to corrosion, and may trap air, gas or sludge (by-products of corrosion), all of which can cause uneven heating or banging noises when in use. The use of a corrosion inhibitor or leak sealer will help to prevent or cure these problems. See also pages 106 and 113. Leaks require prompt action – see page 109.

The header tank ❹
The feed-and-expansion cistern (widely known as the header tank) tops up any water lost from the heating system through leaks or evaporation. The cistern itself may also develop a leak and need replacing. The ballvalve that refills it may jam through lack of use, causing an overflow or allowing air to be drawn into the radiators and pump. A ballvalve can be repaired or replaced (see pages 48-52).

The controls ❺
Controls operating the heating system include a programmer, thermostats and motorised valves. Faulty wiring may lead to malfunction, and mechanical failure may be remedied through servicing or replacement of the faulty control. See page 116 for details on repairing a motorised valve.

HEATING EMERGENCY ACTION

- If the system overheats or the pump fails, turn off the boiler.
- Turn off the gas at the meter immediately if you smell a gas leak and call the National Grid Emergency Service (0800 111 999).
- Until you can make a repair, drain leaking pipes, radiators and hot water cylinders via the relevant drain valve (see page 31).

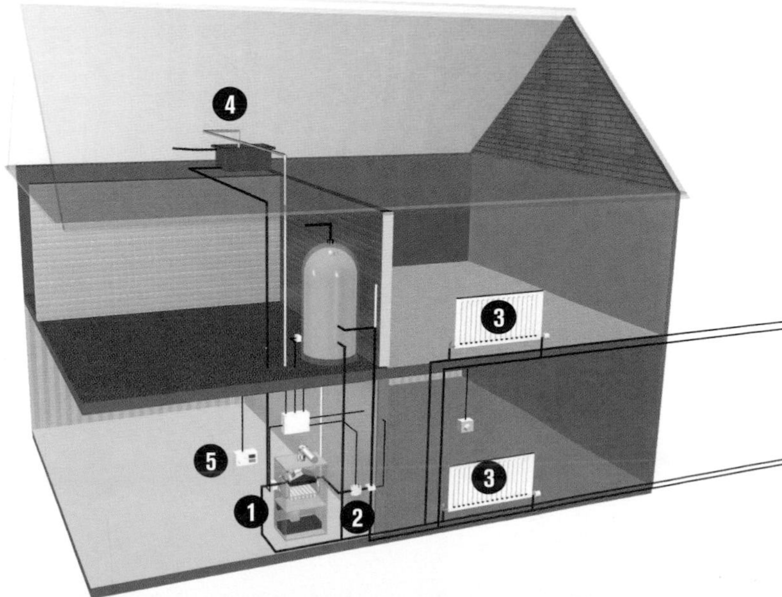

Maintenance and repairs

Tools for plumbing

You will probably have most of the basic tools needed for plumbing – such as spanners and pliers – in your toolkit. For more specialist jobs, tools and equipment can be bought or hired.

Slip-joint pliers

Combination pliers

Pliers A pair of 180mm standard combination pliers is useful for jobs such as removing split pins from ballvalves, and long-nose pliers will grip a sink or washbasin outlet grid. Slip-joint pliers have adjustable jaws and are useful for a wide range of gripping jobs.

OTHER USEFUL TOOLS

Screwdrivers A cross-head (Phillips) screw-driver and two flat-bladed screwdrivers – one large, one small – are sufficient for most plumbing jobs.

Vice and workbench Not essential, but useful for some jobs, such as dismantling a ball valve. Use a grooved jaw lining in an engineers vice. Some portable workbenches incorporate grooves for holding a pipe.

Torch Plumbing is often in dark places. Keep a powerful torch (and spare batteries) handy – preferably one in a square casing so that it will stand up, or one designed to strap to your head.

Ladder Plumbing emergencies usually occur in the loft. A retractable loft ladder is the safest means of access. Alternatively, use a ladder that reaches right into the loft.

Pipe repair clamp For an instant, though temporary, repair to a leaking pipe. The clamp is a two-part collar lined in rubber that you screw together around the pipe.

Plumbing putty An epoxy putty useful for patching slow leaks.

Blowtorch A torch flame is necessary for joining copper pipes with soldered capillary fittings.

Pipe wrenches For gripping pipes, circular fittings or hexagonal nuts that have been rounded off at the edges. Two wrenches are needed for some jobs. Some pipe wrenches, such as Footprint wrenches, are operated by squeezing the handles together, but the Stillson type has an adjuster nut for altering the jaw size. Useful Stillson wrench sizes are 250mm and 360mm, with jaw openings up to 25mm and 38mm. When using a wrench, always push or pull in the direction of the jaw opening. Pad the jaws with cloth if they are likely to damage the fitting – if it is plastic, for example.

PTFE tape Makes threaded fittings watertight. Wrap the tape around the thread before screwing the joint together.

Plunger Inexpensive tool used to clear blockages from a sink, basin or bath. Pump it up and down over the plughole.

Spanners You will need a selection of open-ended spanners for plumbing work. Have at least two crescent-pattern adjustable spanners in different sizes up to 60mm.

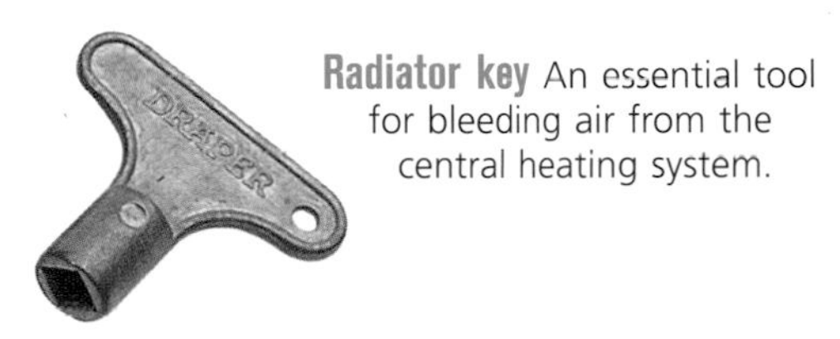

Radiator key An essential tool for bleeding air from the central heating system.

Sink or WC auger Used to dislodge blockages from waste pipework. Some augers have a rotating handle to drive the wire into the blockage and break it up; others work just by pushing the wire manually into the blocked pipe.

TOOLS FOR ADVANCED WORK

Pipe-bending machine A lightweight machine is useful if you have a number of bends to make in a 15mm, 22mm or 28mm diameter pipe, and can be hired. It incorporates a curved former for shaping the bend, and separate guide blocks for each pipe size.

Hacksaw Use a large or small (junior) hacksaw to cut metal or plastic pipes.

Half-round metal file Use for smoothing the burred edge of a cut pipe.

Pipe-bending springs Steel bending springs support the inside of copper pipe while bending it by hand. Springs are sold in 15mm and 22mm sizes.

Basin wrench This wrench (also known as a crowsfoot spanner) is used for tightening or loosening the backnuts of taps.

Wire wool Use to clean inside pipe ends before making joints, and to remove rust.

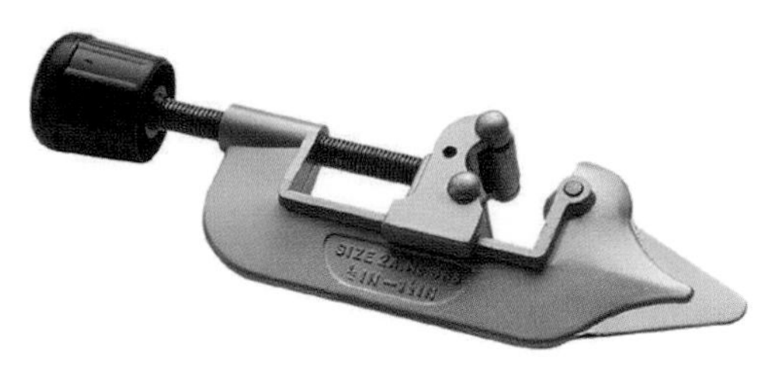

Pipe cutter Quicker and more accurate than a hacksaw for cutting copper pipe. It may have an adjustable slide for cutting tubes of different diameters, and a fixed cutting wheel. Some have a tapered reamer on the front for removing the burr from the inside of the pipe. A small circular cutter known as a pipe slice is useful for cutting pipes in situ. It is fitted round the pipe and rotated to make the cut.

Immersion heater spanner This large spanner is specially designed for removing and fitting the electrical element of an immersion heater. Hire one if you need it.

Cutting off the water supply

In many homes, only the kitchen tap is fed from the rising main; others are fed from the cold water cistern. It depends whether the plumbing system is direct or indirect (page 12).

Taps fed from the cistern

1 To isolate a hot or cold tap supplied from the cistern, turn off the gatevalve on the supply pipe from the cistern. If a service valve (see page 30) is fitted in the pipe to the tap, turn it off with a screwdriver.

2 Turn on the tap until the water has stopped flowing.

Alternatively If there is no gatevalve or service valve on the pipe, you will have to drain the cistern.

Draining the cistern

1 Tie the ballvalve arm to a piece of wood laid across the cistern (see page 31). This stops the flow from the mains.

2 Turn on the bathroom cold taps until the water stops flowing, then turn on the hot taps – very little water will flow from them. (You need not turn off the boiler, as the hot water cylinder will not be drained.)

Taps fed from the rising main

Turn off the main indoor stoptap, then turn on the mains-fed tap until the water stops.

Draining the rising main

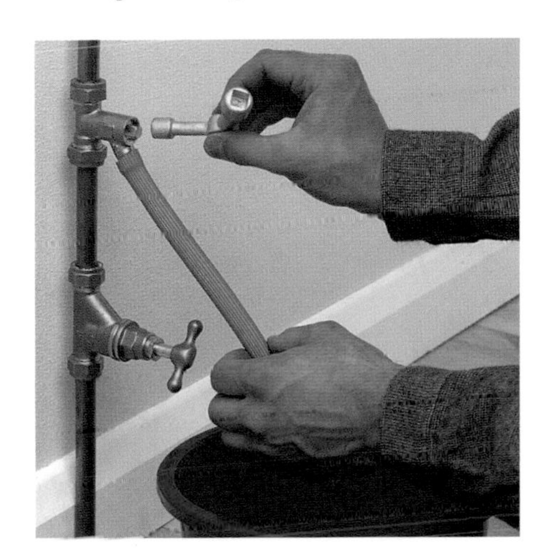

You may want to drain the rising main to take a branch pipe from it or to repair the main stoptap. Find the drain valve above the stoptap, fit a short piece of hose to its outlet and open it with a drain valve key or pliers. Catch the water, usually only a litre or two, in a bucket.

> **HELPFUL TIP**
>
> A stoptap that has been open for a long time may be jammed. To guard against this, close and open the stoptap fully twice a year. After opening it, give the handle a quarter turn towards closure. This prevents jamming without affecting water flow. If the indoor stoptap is difficult to turn, don't try to force it. Isolate it by turning off the stoptap outside the house. Then the indoor one can be dismantled and serviced.

Turning off the outdoor stoptap

You may need to turn off the outdoor stoptap if the indoor one is broken, jammed or has a leak from the spindle. Stoptap keys can be bought from plumbers' merchants, but first check the type needed – the tap may have a crutch handle (see right) or a square spindle.

Alternatively If you have no stoptap key, make your own. Take a piece of strong wood about 1m long and in one end cut a V-shaped slot about 25mm wide at the opening and 75mm deep. Securely fix a piece of wood as a cross-bar handle at the other end. Slip the slot over the stoptap handle to turn it. This tool will not turn a stoptap with a square spindle.

1 Locate the stoptap, which is under a cover, about 100mm across, just inside or just outside the boundary of your property. If you cannot find the outdoor stoptap, call your water supply company.

2 Raise the cover. This may be difficult if it has not been raised for some time.

3 Insert the stoptap key into the guard pipe and engage the stoptap handle at the bottom. Turn it clockwise to close.

TYPES OF STOPTAP AND ISOLATING VALVE

Stoptap A tap with a valve and washer that is inserted into a mains-pressure supply pipe to control the water flow through it. A stoptap is usually kept turned on, being turned off only when necessary to cut off the supply. It must be fitted the right way round (an arrow mark shows the flow direction). Most stoptaps have a crutch handle.

Drain valve A tap without a handle, opened by turning the spindle with a drain valve key. It is normally kept closed, but has a ribbed outlet for attaching a hose when draining is necessary. A drain valve is fitted in those parts of the plumbing system that cannot be drained through household taps – for instance, in the boiler or central-heating systems and on the rising main.

Gate valve An isolating valve with a wheel handle, through which the water flow is controlled by raising or lowering a metal plate (or gate). It can be fitted either way round and is normally used in low-pressure pipes such as supply pipes from a storage cistern. With the gate open, the flow is completely unrestricted. When it is closed, the seal is not as watertight as a stoptap.

Service valve A small isolating valve operated with a screwdriver. This turns a pierced plug inside the valve to stop or restore the water flow. Normally used in a low-pressure supply pipe to a tap or ballvalve to cut off the water for repairs. A similar valve with a small lever handle and a threaded outlet is used to control the flow to the flexible supply hoses of a washing machine or dishwasher.

Draining down the heating system

Central heating systems sometimes have to be drained down – to repair a leak, for example. The following method is for an open-vented system, the most common type.

1 Switch off the boiler at the programmer. After a few minutes switch off the main electricity supply to the heating system. This is usually a fused spur located near the programmer.

2 If there is a solid fuel fire serving as a boiler, make sure the fire is out and the boiler is cold.

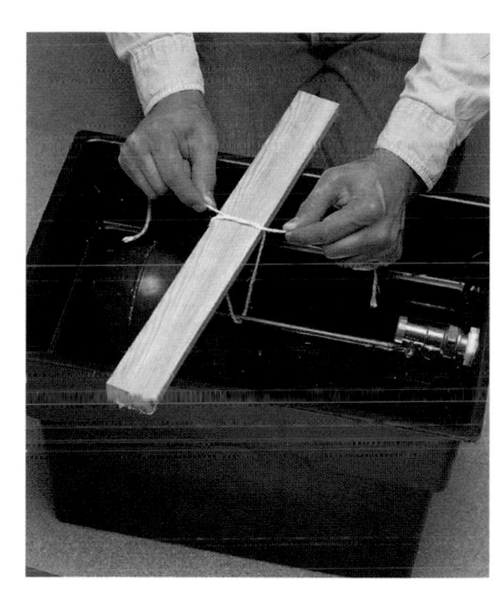

3 Shut off the water supply to the feed-and-expansion cistern. There should be a separate stoptap for this on the branch pipe from the rising main connected to the cistern's ballvalve. If there is no separate stoptap, or it is jammed and cannot be turned, stop the water flow into the cistern by tying up the ballvalve to a piece of wood laid across the top of the cistern.

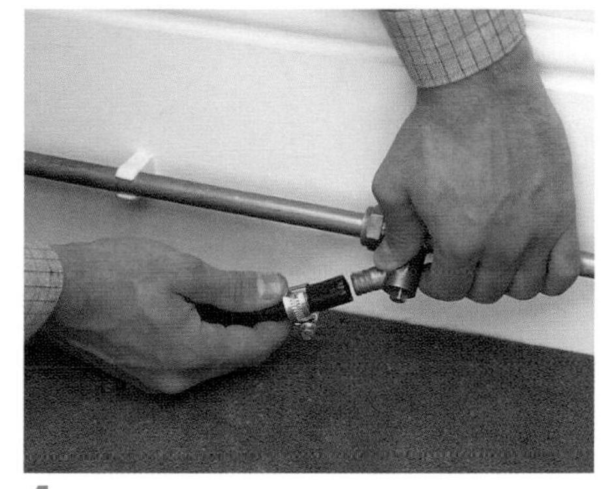

4 Locate the drain valve, which may be near the bottom of the boiler. There may be more than one drainage point on the system. Clip a garden hose onto the outlet and run the hose to a drain outside. Open the drain valve so the water starts to flow.

5 Locate all the points at which air is vented from the central heating system. There will be radiator vents, a vent on the primary flow near the hot water cylinder in fully pumped systems, and manual or automatic vents in the loft if circulating pipes run there. There could be additional vents at other points as well.

6 Open the drain valve with a spanner, pliers or an Isle of Man key (shown above), turning counter-clockwise. Water will then start to flow out of the hose at a fairly slow rate.

7 Start opening the venting points at the top of the system. This will greatly speed up the flow from the drain valve. As the water level drops further, open the lower venting points until they are all open.

Refilling the system

1 Close all the drain valves and all the air vents in the system. Then check that all work on the system is finished.

2 Turn on the stoptap to the header tank, or untie the ballvalve, in order to let water back into the system.

3 Open one of the lowest air vents until water starts to flow out, then close it. Repeat with the lower air vents until the bottom of the central heating system is full of water. Then do the upper vents, and close them when the system is full.

4 Make sure that the ballvalve to the header tank has closed. The water level in the cistern should be around 100mm above the outlet: the rest of the cistern space is to take up the expansion of the water in the system as it heats up.

5 If the water level is too high, close off the mains water supply to the cistern and open the drain valve to let some out.

COMBINATION BOILER OR SEALED SYSTEMS

If you have a combination boiler or a boiler running with a sealed system (see page 19), the water in the system is 'topped up' by a special flexible filling loop connected to the rising main. This is fitted with control valves and is normally connected (to the heating return pipe) only when needed. A pressure gauge is fitted so that you know how much water to let in.

Adjust the arm on the ballvalve so that it closes the valve at the correct water level. Check that the cistern's lid and insulating jacket are in place.

6 Switch on the electricity and turn on the gas. Re-light the pilot light in a gas boiler. Turn on the system at the programmer or timeswitch. Turn up the room thermostat.

7 Re-light the boiler, following the manufacturer's instructions.

8 As the system heats up, more venting will be necessary in order to release air driven off from the water. Minor venting will be required for a few days.

9 Check for leaks.

HELPFUL TIP

When you open a drain valve it is always a good idea to check the condition of the washer. If it looks worn or perished, replace it. While the system is drained down it is also an excellent time to fit isolating service valves to various sections to save you draining down the whole system in the future.

If your system contained corrosion inhibitor, this needs to be replaced.

Repairing a burst pipe

Metal pipes are more likely to suffer frost damage than plastic pipes. Copper and stainless steel pipes are less vulnerable than softer lead pipes.

As an ice plug forms, it expands and may split the pipe or force open a joint. Then when the ice melts, the pipe or fitting leaks.

A split copper or plastic pipe can be temporarily repaired with a proprietary burst-pipe repair clamp (opposite, top). Cut off the water supply (page 29), drain the pipe and replace the damaged length. For a split less than 90mm long in a copper pipe, you can make a permanent repair with a slip coupling. For lead piping, use a tape-repair kit for a strong repair that will allow you to restore the water supply until a plumber can make a permanent repair (working on lead pipework is best left to a professional).

Using a slip coupling

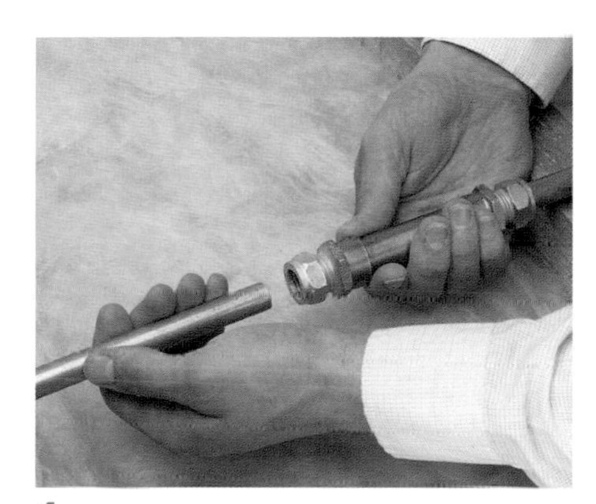

1 Cut out the damaged part and slide the slip end of the coupling (with no pipe stop) onto a pipe end. Then push it onto the other end. If it will not go in, unscrew the backnuts and slide the nuts and olives at each end along the pipe first.

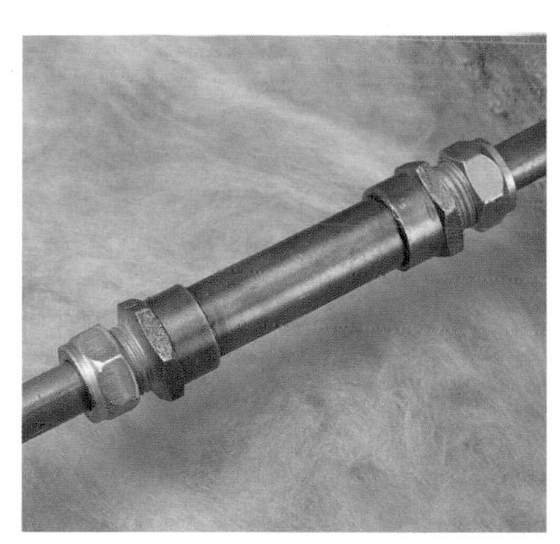

2 Refit the nuts and olives and screw them up hand-tight. Then tighten the nuts for one and a quarter turns with a spanner.

Alternatively Fit an emergency pipe repair clamp and tighten the screws fully with a screwdriver.

Compression tape

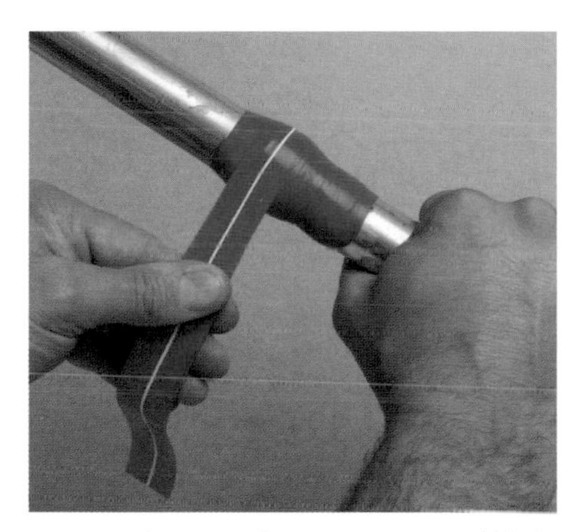

Compression tape, known as LLFA or 'Alta' tape, is a quick way of repairing any hot or cold water pipe. The tape forms a permanent repair and can withstand high pressure and heat.

Lead pipe emergency

If you have a burst on a lead pipe and can't turn off the water, flatten the pipe with a hammer to stem the flood of water. The pipe can be replaced with plastic or copper at a later date, using adaptors.

Dealing with an airlock

If the water flow from a tap (usually a hot tap) is poor when fully turned on, then hisses and bubbles and stops altogether, there is an airlock in the supply pipe.

Tools *Length of hose with a push-fit tap adaptor at each end. Possibly also dishcloth; screwdriver.*

1 Connect one end of the hose to the tap giving the trouble. If it is a bath tap and the hose is difficult to fit, connect to the nearby washbasin hot tap instead.

2 Connect the other end of the hose to the kitchen cold tap or to another mains-fed tap. Turn on the faulty tap first, then the mains-fed tap. The pressure of the mains water should blow the air bubble out of the pipe.

Important When applying mains pressure to a pipe in a stored-water circuit, there is a very slight risk of water from the system contaminating the mains water supply. Therefore you should do the job quickly and disconnect the hose immediately afterwards.

Airlock in a kitchen mixer

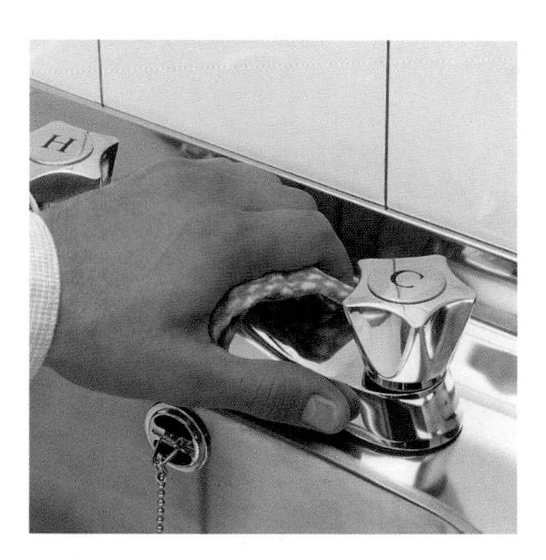

If the hot tap will not work, remove the swivel spout (page 41) and hold a cloth firmly over the spout hole while you turn on first the hot tap then the cold tap.

If airlocks keep occurring

There are many ways that air can be drawn into the water system and cause airlocks. Check these possibilities:

- Is the cold water cistern too small for the household's needs? If it is smaller than the standard 230 litres, replace it with one of standard size.

- Is the ballvalve in the cold water cistern sluggish? Watch the cistern emptying while the bath fills. If the valve does not open wide enough as water is drawn off, there will be a slow inflow and the cistern will empty before the bath is filled, allowing air to be drawn into the supply pipe. Dismantle and clean the ballvalve (page 48).

- Is the supply pipe from the cold water cistern to the hot water cylinder obstructed or too narrow? Check that any gatevalve is fully open, and replace the pipe if it is narrower than 22mm diameter. If hot water drawn for a bath is not replaced quickly enough, the water level in the vent pipe will fall below the level of the hot water supply pipe, and air will enter.

Dealing with a blocked sink

Grease may have built up in the trap and waste pipe, trapping food particles and other debris, or an object may be obstructing the waste pipe.

Tools *Possibly a length of wire; sink-waste plunger; sink auger or a length of expanding curtain wire; bucket.*

Materials *Caustic soda or proprietary chemical or enzyme cleaner; petroleum jelly.*

Sink slow to empty

If a sink is slow to empty, smear petroleum jelly on the rim of the plug hole to protect it, and then apply proprietary chemical or enzyme cleaner according to the manufacturer's instructions.

Sink completely blocked

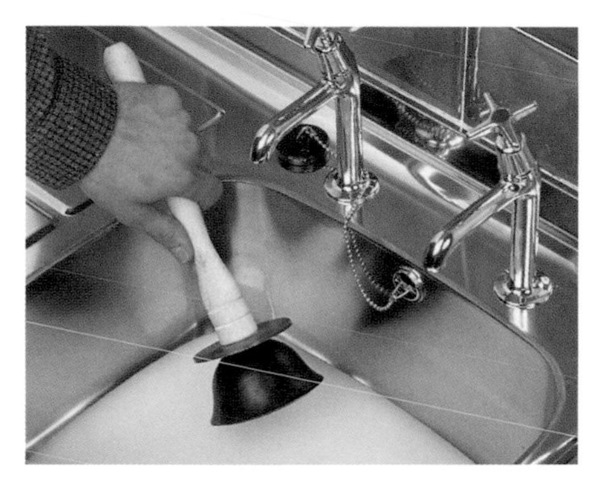

1 If the water will not run away at all, place the sink plunger cup squarely over the plug hole.

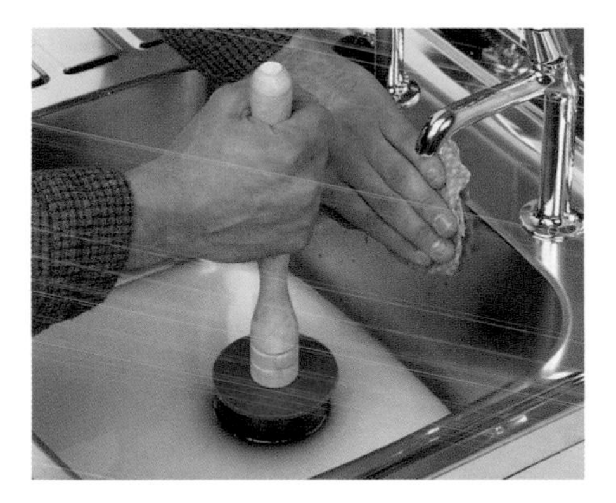

2 Stuff a damp cloth firmly into the overflow opening and hold it there. This stops air escaping through the hole and dissipating the force you build up by plunging. Pump the plunger sharply up and down. If the blockage does not clear, repeat the operation.

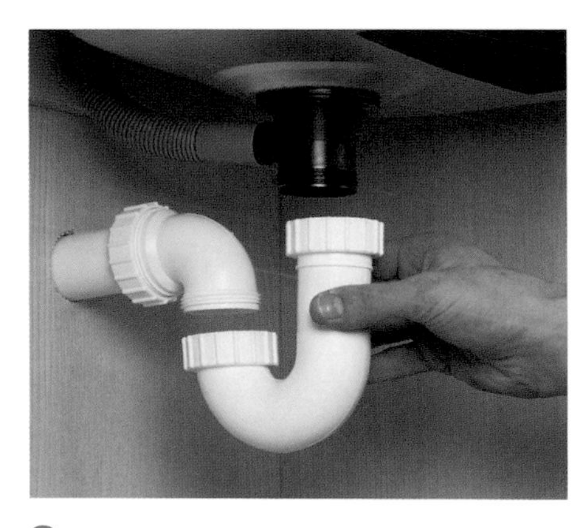

3 If plunging fails, replace the sink plug. Put a bucket under the sink and disconnect the trap. Wash it out thoroughly if it is blocked with debris.

4 If the obstruction is not in the trap, try using a sink auger. It is a spiral device that can be hired or bought. Disconnect the blocked pipe from its trap and feed the wire into it. Then turn the handle to rotate the spiral. This drives its cutting head into the blockage and breaks it up.

Alternatively If you have a vacuum cleaner that is designed to cope with liquids, you can use it to try to dislodge a blockage in a sink trap. Press a cloth over the overflow in the sink. Then place the suction tube of the vacuum over the plughole and switch on. This will probably loosen the blockage sufficiently to allow it to be carried away by the water flow through the trap.

Alternatively If you have poured fat into the sink and it has hardened, try warming the pipe with a hair dryer, to melt the grease. Flush plenty of hot water after it.

Other pipe blockages

The waste pipes from washing machines and dishwashers are often connected to the under-sink waste trap. Alternatively, they may join the main waste pipe at a T-junction away from the sink. If all your appliances feed into the one trap, you may need to disconnect all the pipes in turn and then clean each one to clear a blockage.

If cleaning the trap and waste pipe does not work, the blockage may be in the gully outside or in the underground drains. Blocked gullies can usually be cleared by hand; blocked underground drains will need at least some drain rods (which you can hire). If drains are seriously blocked (perhaps by tree roots growing into them), you may need to call in a professional drain clearance company.

How taps work

All taps work in much the same way – a rotating handle opens and closes a valve inside the body of the tap. Traditional taps use screw threads to move a rubber washer away from a seat; modern taps have two slotted ceramic discs.

Non-rising spindle As the shrouded head is turned, the spindle lifts the 'jumper' to which the washer is fitted. Leaks past the spindle are prevented by a rubber O-ring and the handle and headgear have to be removed to replace the washer.

Ceramic disc tap In this type of tap, one ceramic disc is rotated against another until openings in the discs line up and water can flow through. Only a quarter turn is needed.

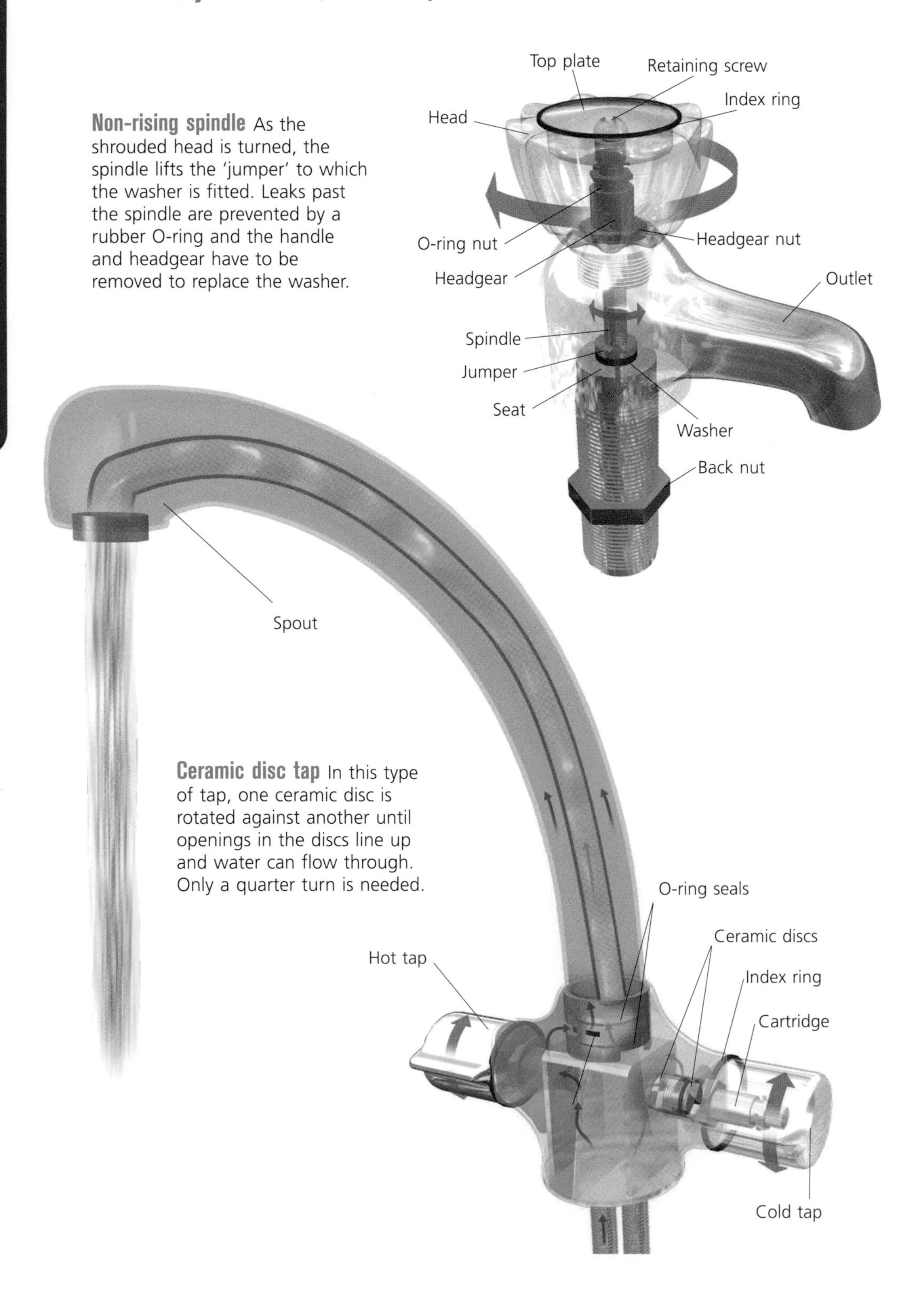

Repairing a dripping tap

A dripping tap usually means that the tap washer needs renewing, but can also be caused by a damaged valve seating. If the drip is from a mixer spout, renew both tap washers.

Tools *One large open-ended spanner, normally 20mm for a 12mm tap or 24mm for a 19mm tap (or use an adjustable spanner); old screwdriver (for prising). Possibly also one small spanner (normally 8mm); one or two pipe wrenches; cloth for padding jaws; one 5mm, one 10mm screwdriver.*

Materials *Replacement washer or a washer-and-jumper valve unit; alternatively, a washer-and-seating set; petroleum jelly. Possibly also penetrating oil.*

Rising spindle Here, the spindle of the tap is directly attached to the jumper/washer assembly and the whole lot moves up when the capstan head is turned. The rising spindle is sealed with packing: to replace the washer, the handle, metal shroud and the headgear all have to be removed.

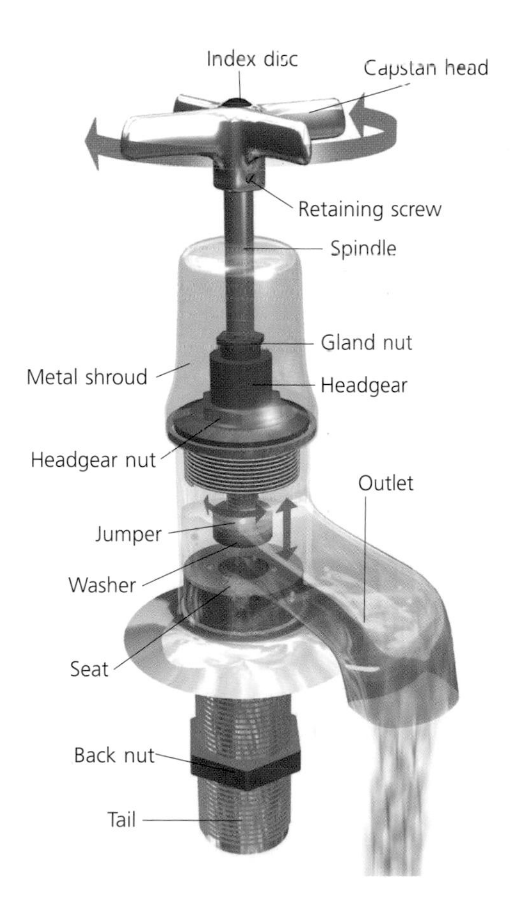

Removing the headgear

1 Cut off the water supply (page 29). Make sure the tap is turned fully on, and put the plug into the plughole to stop any small parts falling down the waste pipe.

2 Unscrew or lever off the cover of a non-rising spindle tap to expose the retaining screw. Remove the screw and put it in a safe place. Remove the head.

Alternatively With a rising spindle tap, prise off the index disc (if necessary) and remove the retaining screw to release the capstan head from the spindle. Use a wrench wrapped in cloth to unscrew the metal shroud and lift it away from the headgear nut.

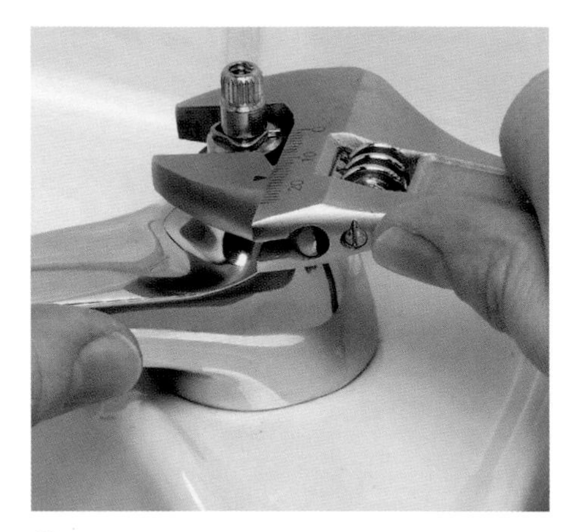

3 Undo the headgear nut with a spanner. Do not force the nut, if it is stiff. Brace the tap body by hand or with a pipe wrench wrapped in a cloth, to prevent the tap from turning and fracturing a ceramic basin or bidet.

4 If the nut is still difficult to turn, apply penetrating oil round the joint, wait about 10 minutes to give it time to soak in, then try again. You may have to make several applications.

Fitting the washer

1 Prise off the washer with a screwdriver. If there is a small nut holding it in place, unscrew it with a spanner (normally 8mm). If it is difficult to undo, put penetrating oil round it and try again when it has soaked in. Then prise off the washer.

Alternatively If the nut is impossible to remove, you can replace both the jumper valve and washer in one unit.

2 After fitting a new washer or washer and jumper, grease the threads on the base of the tap before reassembling.

Repairing the valve seating

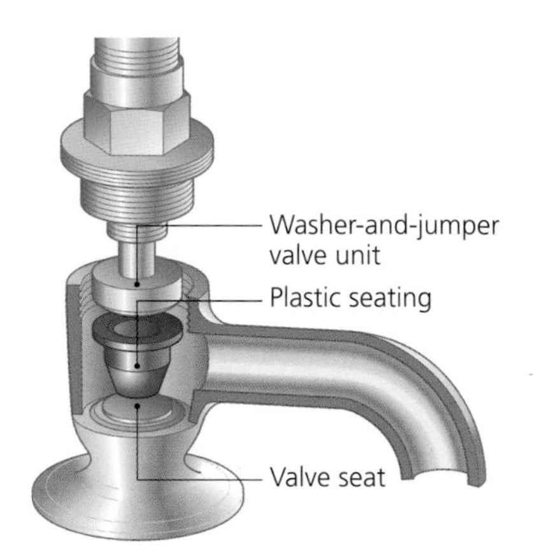

When renewing a washer, inspect the valve seat inside the tap body. If it is scaled or scored by grit, the seal between washer and seat will not be effective even with a new washer.

The simplest repair is with a washer-and-seating set. This has a plastic seat to fit into the valve seat, and a washer-and-jumper valve unit to fit into the headgear.

When the tap is turned off, the plastic seating is forced firmly into place. It may take a few days for the new seating to give a completely watertight fit.

An alternative repair is to buy or hire a tap reseating tool and grind the seat smooth yourself.

AVOIDING HARD-WATER DAMAGE TO TAPS

If you live in a hard-water area, check your taps for damage, once a year.

Turn off the mains water supply. One at a time check that the headgear on each tap unscrews easily. Use penetrating oil to release stiff nuts and use a spanner and a wrench wrapped in a cloth to hold the body of the tap as you turn.

If limescale has built up, remove and soak small parts in vinegar or limescale remover. Smear the thread with lubricant before reassembling.

Tap conversion kits

You may be able to get tap conversion kits to change the style of taps and replace worn or broken mechanisms. Newer heads can be changed back to Victorian brass heads, or a tap with a crutch or capstan handle can be given a newer look. The spout and body of the tap remain in place.

Some kits have bushes to fit different tap sizes. The kits are available from most DIY stores and fitting instructions are included.

Cleaning or replacing ceramic discs

Ceramic disc taps operate on a different principle from conventional taps that have washers and spindles. Positioned in the body of the tap is a cartridge containing a pair of ceramic discs, each with two holes in it.

One disc is fixed in position; the other rotates when the handle is turned. As the movable disc rotates, the holes in it line up with the holes in the fixed one and water flows through them. When the tap is turned off the movable disc rotates so that the holes no longer align. Ceramic disc taps are easier to use than conventional taps and last longer before needing repair. On the other hand, they are more expensive and when they do leak, need to be repaired quickly as they can gush rather than drip. Have replacement cartridges ready to hand.

Dealing with a dripping tap

If a scratched ceramic disc is causing the leak, the entire cartridge must be replaced: left-handed for a hot tap or right-handed for a cold tap. Remove the old cartridge and take it with you when buying a replacement to make sure it is the correct size and 'hand'. Mixer taps can also drip at the base of the spout if the O-ring seal has perished. Replace if necessary.

Checking discs in a ceramic disc mixer tap

1 Turn off the water supply. Pull off the tap handles (it may be necessary to unscrew a small retaining screw on each) and use a spanner to unscrew the headgear section.

2 Carefully remove the ceramic cartridges, keeping hot and cold separate. Check both cartridges for dirt and wear and tear.

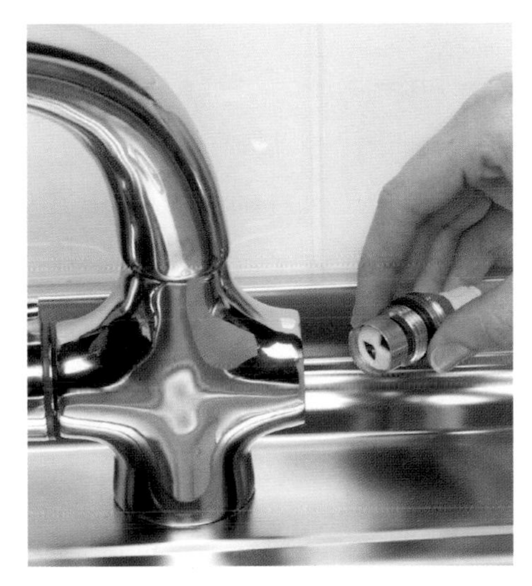

3 If the cartridges are worn, replace with identical parts for the tap unit. Make sure the hot and cold cartridges are fitted into the correct taps.

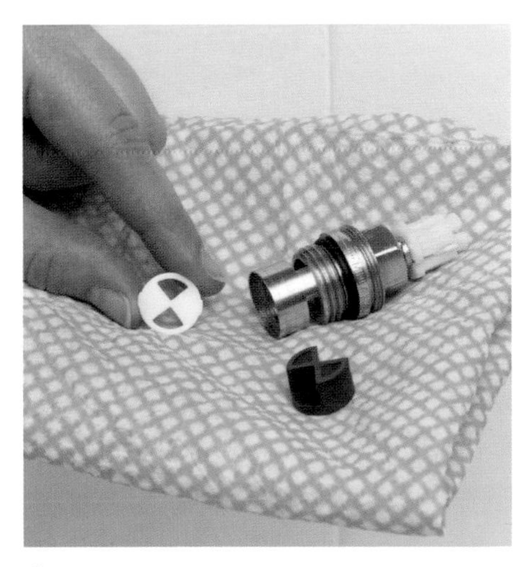

4 If the cartridges are dirty, clean them with a damp cloth. Replace the rubber seal, if it is worn. Replace the cartridge in the tap unit, fitting the hot and cold cartridges into the appropriate taps.

Curing a leak from a spindle or spout

Leakage from the body of the tap – from round the spindle, the base of a swivel spout, or the diverter knob on a shower mixer tap – may indicate a faulty gland or O-ring seal.

Possible causes This sort of leak is most likely to occur on a kitchen cold tap with a bell-shaped cover and visible spindle. Soapy water from wet hands may have run down the spindle and washed the grease out of the gland that makes a watertight joint round the spindle. If the tap is used with a garden hose, back pressure from the hose connection will also weaken the gland.

On a modern tap, especially one with a shrouded head, there is an O-ring seal instead of a gland, and it rarely needs replacing. However, an O-ring seal may occasionally become worn.

Tools *Small spanner (normally 12mm) or adjustable spanner. Possibly also one 5mm and one 10mm screwdriver; penknife or screwdriver for prising; two small wooden blocks about 10mm deep (such as spring clothes pegs).*

Materials *Packing materials (gland-packing string or PTFE tape). Possibly also silicone grease; O-rings (and possibly washers) of the correct size – take the old ones with you when buying, or give the make of tap.*

RELEASING THE SPINDLE

A non-rising spindle tap may have a circlip keeping the spindle in place. When you have removed the headgear, lever out the circlip so that you can gain access to the worn O-rings.

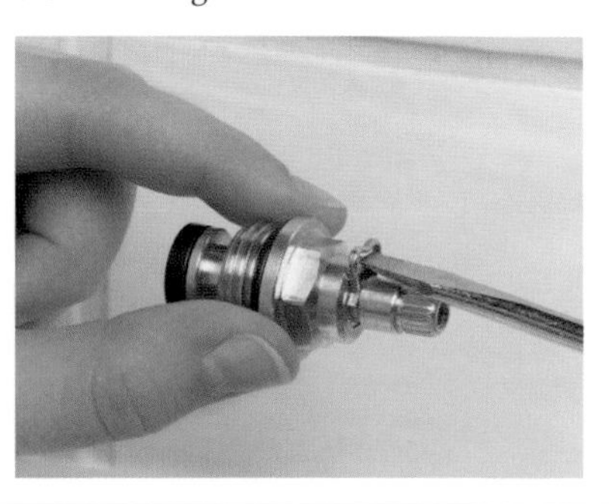

Adjusting the gland

There is no need to cut off the water supply to the tap.

1 With the tap turned off, undo the small screw that secures the capstan handle and put it in a safe place (it is very easily lost), then remove the handle. If there is no screw, the handle should pull off.

2 Remove the bell-shaped cover to reveal the gland nut – the highest nut on the spindle. Tighten the nut about half a turn with a spanner.

3 Turn the tap on by temporarily slipping the handle back on, then check whether there is still a leak from the spindle. If there is not, turn the gland nut another quarter turn and reassemble the tap. Do not overtighten the gland nut, or the tap will be hard to turn off.

4 If there is still a leak, give another half turn and check again.

5 If the gland continues leaking after you have adjusted it as far as possible, repack the gland.

Replacing the packing

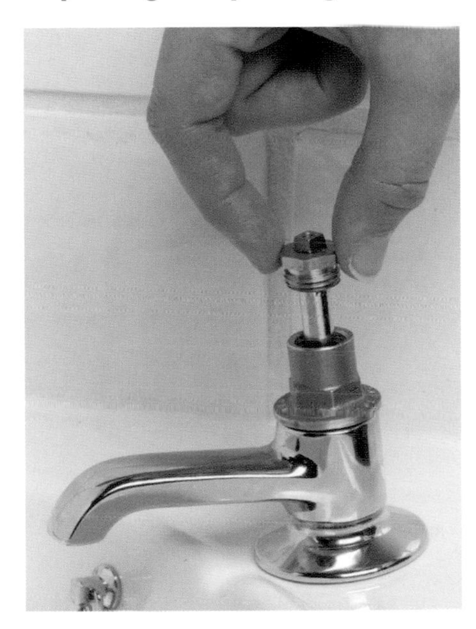

1 With the tap turned off and the handle and cover removed, use a spanner to remove the gland nut and lift it out.

2 Pick out the old packing with a small screwdriver. Replace it with packing string from a plumbers' merchant or with PTFE tape pulled into a thin string. Pack it in with a screwdriver, then replace the gland nut and reassemble the tap.

Renewing the O-ring on a shrouded-head tap

1 Cut off the water supply to the tap (page 29) and remove the tap handle and headgear in the same way as for renewing a washer.

2 Hold the headgear between your fingers and turn the spindle clockwise to unscrew and remove the washer unit.

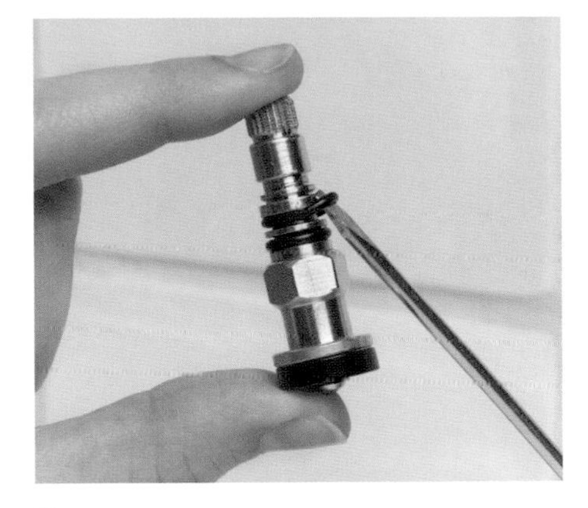

3 Prise out the O-ring at the top of the washer unit with a screwdriver or penknife.

4 Smear the new O-ring with silicone grease, fit it in position, and reassemble the tap.

Renewing O-rings on a kitchen mixer tap

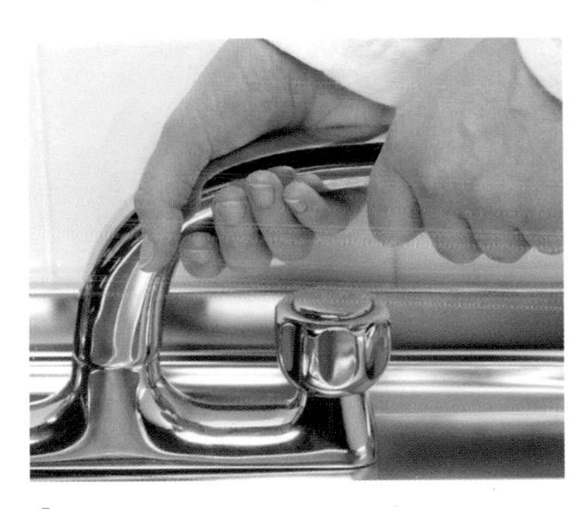

1 With both taps turned off, remove any retaining screw found behind the spout. If there is no screw, turn the spout to line up with the tap body and pull upwards sharply.

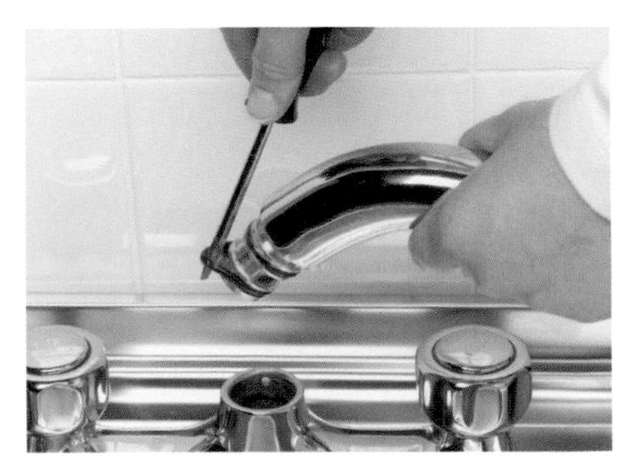

2 Note the position of the O-rings (probably two) and remove them.

REPACKING A STOPTAP WITH FIBRE STRING

The gland on a capstan-handle stoptap is the type most likely to need repacking. Use fibre string (from a plumbers' merchant) or PTFE tape.

1 Turn off the stoptap. Undo the gland nut, slide it up the spindle and remove it.

2 Rake out the gland packing with a penknife or similar tool.

3 To repack the gland with fibre string, steep a length in petroleum jelly and wind and stuff it into the gland with a screwdriver blade. Wind and push the string in until it is caulked down hard, then reassemble the tap.

3 Coat new O-rings of the correct size with silicone grease and fit them in position.

4 Smear a little petroleum jelly onto the end of the spout and then refit it to the tap body.

Replacing shower-diverter O-rings

Diverters vary in design, but most have a sprung rod and plate attached to the diverter knob. When the knob is lifted, the plate opens the shower outlet and seals the tap outlet for as long as the shower is on.

1 With the bath taps turned off, lift the shower-diverter knob and undo the headgear nut with a spanner (probably 12mm size or use an adjustable spanner).

2 Lift out the diverter body and note the position of the washers and O-rings.

3 Remove the knob from the diverter body by turning it counter-clockwise. You may need to grip it with a wrench.

4 Withdraw the rod and plate from the diverter body and remove the small O-ring at the top of the rod.

5 Grease a new O-ring of the correct size with silicone grease and fit it in place.

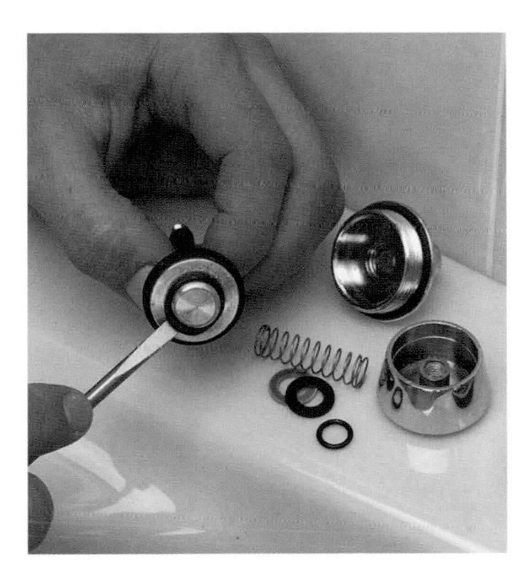

6 Replace all other rubber washers and O-rings on the base of the rod and plate. Old ones may have to be prised out.

Repairing a faulty WC cistern

An overflow or a failure of a WC to flush properly are often caused by a faulty ballvalve, which governs the water level in the cistern.

Before you start Failure to flush properly is caused by either a low water level, or a worn or damaged flap valve. To determine the cause, first check the water level.

The plastic lever arm linking the spindle of the flush control to the siphon lift rod may eventually wear out and break. Replacement lever arms can be bought and are easily fitted.

Checking the water level to find out if the ballvalve is at fault

1 Remove the cistern lid (it may lift off or be held by one or more screws). When the cistern is full, the water level should be about 25mm below the overflow outlet. Or there may be a water level marked on the inside wall of the cistern.

2 If the level is above this and the cistern is overflowing, either through the external pipe or into the bowl, turn off the water supply, flush the WC, then repair or adjust the ballvalve.

Continued on page 44

FLUSHING PROBLEMS

The most common problems with WCs are that the WC will not flush, that water runs continuously into the pan or that water runs continuously into the cistern and out through the overflow pipe. Most of these problems are easily fixed.

WC will not flush Check that the flushing lever is still attached to the internal workings of the cistern. Reattach the link or improvise a replacement from a length of thick wire. If the link is still in place, the flap valve may need replacing – this is often the cause if you need to operate the lever several times before the WC will flush.

Continuous flushing When water keeps running into the pan, the siphon may have split or the sealing washer at the base of the siphon may have perished. Both can be replaced. Alternatively, the cistern may be filling too fast, so that the siphoning action of the flush mechanism cannot be interrupted. Fit a restrictor in the float valve to reduce the water flow.

Overflowing cistern Continuous filling may be caused by a faulty ballvalve or a badly adjusted float arm.Try adjusting the float arm before you replace the valve.

Renewing the flap valve in a standard siphon

The flap valve for a standard siphon may be sold under the names siphon washer or cistern diaphragm. If you do not know the size you want, buy the largest available and cut it down.

Tools *Screwdriver; wooden batten slightly longer than the cistern width; string; pipe wrench with a jaw opening of about 65mm; bowl or bucket. Possibly also sharp coloured pencil; scissors; container for bailing, or tube for siphoning.*

Materials *Plastic flap valve. Possibly also O-ring for ring-seal joint.*

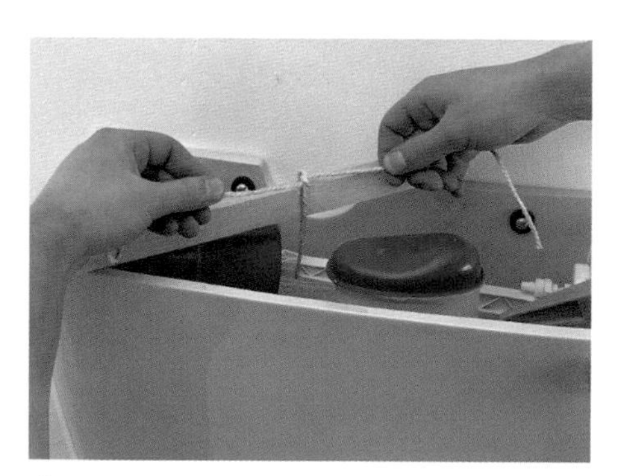

1 Turn off the water. If you can't do this easily, then hold the ballvalve up with string and a length of wood.

2 Empty the cistern. If it cannot be flushed at all, bail or siphon the water out.

3 Use a large pipe wrench to undo the lower of the two large nuts underneath the cistern, then disconnect the flush pipe and push it to one side.

4 Put a bowl or bucket underneath the cistern and undo the large nut immediately under the cistern (this is the siphon-retaining nut). Water will flow out as you loosen the nut.

5 Unhook the lift rod from the flushing lever, lift the inverted U-pipe (the siphon) out from the cistern and lay it on its side.

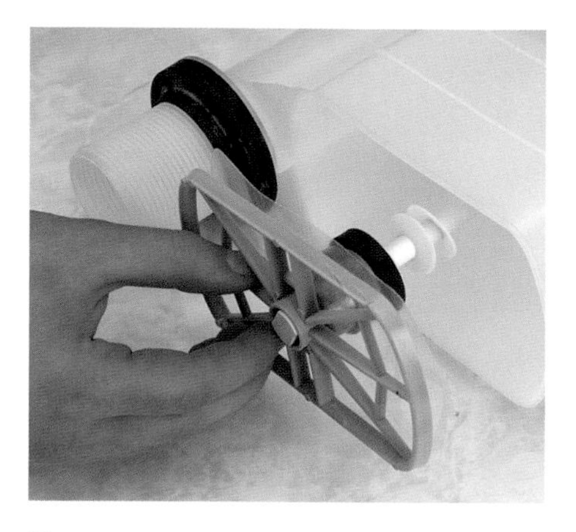

6 Pull out the lift rod and plate and remove the worn flap valve. If the new valve is too big, cut it down with scissors using the old valve as a pattern. It should touch, but not drag on, the dome sides.

7 Fit the new valve over the lift rod and onto the plate, then reassemble the flushing mechanism and reconnect the cistern.

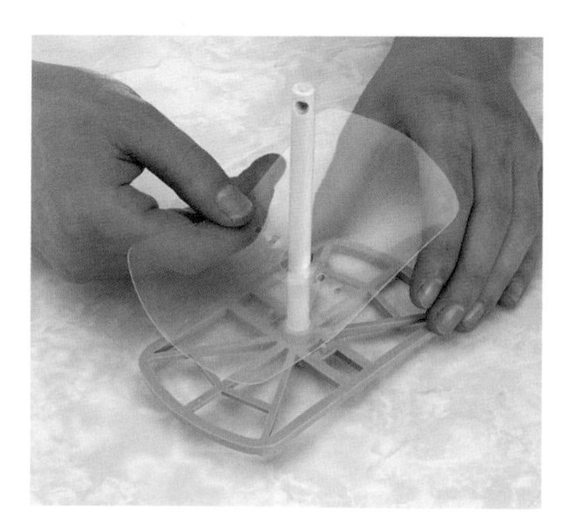

Renewing the flap valve on a close-coupled suite

On some close-coupled suites, the siphon is held by two or more bolts inside the cistern rather than by a large nut underneath. Except for this difference, the flap valve is renewed in the same way as on a standard suite.

On others, the cistern must be lifted off in order to disconnect the siphon. The flap valve can then be renewed in the same way as on a standard suite. Lift off the cistern as follows:

1 Cut off the water supply to the cistern in the same way as for a tap. Empty the cistern by flushing, bailing or siphoning out the water.

2 Disconnect the overflow pipe and water supply pipe from the cistern. They generally have screw fittings with a back nut.

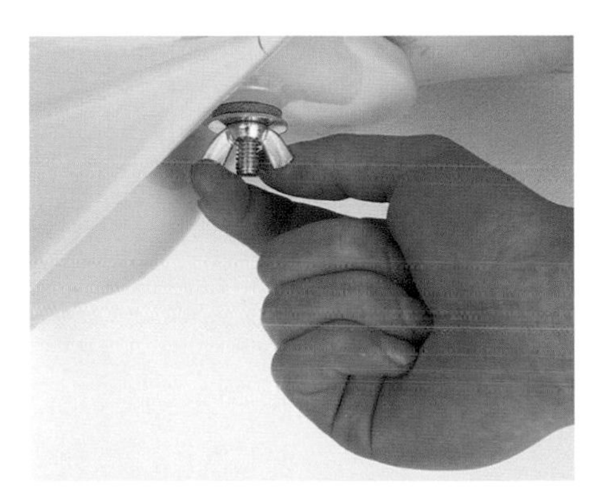

3 Undo the screws holding the cistern to the wall, and the wing nuts securing it to the rear platform of the pan.

4 Lift off the cistern from the pan and unhook the lift rod. Turn the cistern over, unscrew the retaining nut, remove the siphon and plate and renew the valve as opposite.

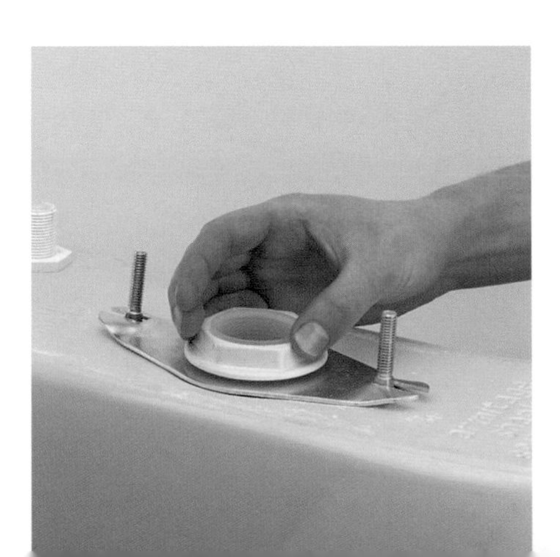

Renewing the flap valve on a two-part siphon

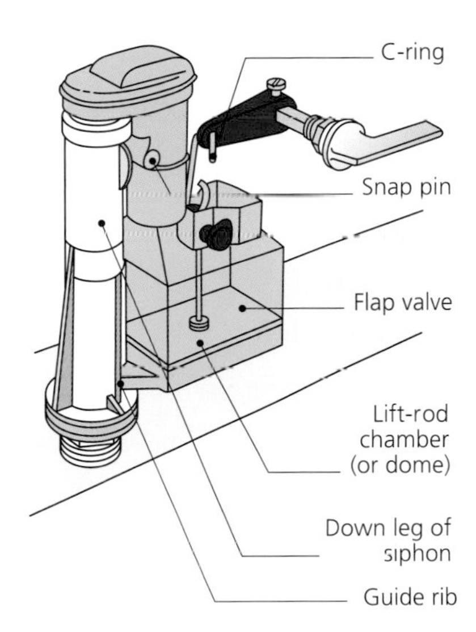

If a cistern is fitted with a two part plastic siphon, there is no need to stop the inflow or, with a close-coupled suite, to remove the cistern.

A two-part siphon can be fitted to most types of WC cistern. The initial fitting does involve cutting off the water supply and, if necessary, lifting off the cistern (see left). After that, maintenance is as below.

Tools *Screwdriver.*

Materials *Spares pack for size of siphon (containing flap valve); washers. Possibly also O-ring-seal.*

1 With the cistern lid removed, unhook the flush lever from the lift-rod C-ring. Remove a lever-type flush handle, as it may be in the way later.

2 Withdraw the snap pin about 30mm to disconnect the lift-rod chamber from the down leg of the siphon.

3 Slide the chamber upwards to disengage it from the guide rib on the down leg.

4 Remove the C-ring and washer from the top of the lift rod and slide the lift rod from the bottom of the chamber.

5 Take off the lift-rod washers and weight so that you can remove the old flap valve and fit a new one.

6 Before reassembling, check if the O-ring seal at the top of the chamber section is worn. Renew it if necessary.

HOW A PUSH-BUTTON FLUSH WORKS

Many modern slimline WC cisterns are too small to accommodate a traditional ball float-operated inlet valve and siphon flush mechanism, operated by a lever and float arm.

Instead, the inlet valve is either a modified diaphragm type with a very short float arm and miniature float, or an ingenious vertical valve with a float cup that fits round the central column of the valve body. Both are very quiet in operation, although the float-cup valve can be slow to refill the cistern if it is supplied with water from a storage cistern, rather than being plumbed in directly to the mains.

In these slimline mechanisms, the traditional siphon flushing method (page 76) is replaced by a plastic valve-operated flush mechanism that is activated by a top-mounted push button in the cistern lid. The mechanism also incorporates an integral overflow. The push button is in two parts: you depress one part for a low-volume flush, and both sections of the button for a full-volume flush. The push button is linked to a plunger to operate the flush, rather than the conventional wire link and lift rod of a traditional flushing mechanism.

Repair and maintenance

The float-cup inlet valve contains a rubber ring seal at the base of the cistern, which may need replacing in time, and these may not be widely available. Be sure to keep the installation instructions after you have installed the valve so that you have a record of where to obtain any spare parts in the future.

How a ballvalve works

In a cold water storage cistern or WC cistern, the water level is regulated by a ballvalve that is opened and closed by a lever arm attached to a float.

Understanding the system

A cold water storage cistern, or a WC cistern where the supply is direct from the mains, needs a high-pressure valve. A WC cistern supplied from the cold water storage cistern needs a low-pressure valve. Check the packaging when you buy as the two versions look similar. Equilibrium valves can be adapted for high or low pressure.

Low-pressure valves have wider inlet nozzles than high-pressure valves. If a high-pressure valve is fitted where a low-pressure valve is needed, the cistern will fill much too slowly. If a low-pressure valve is fitted in a cistern supplied from the mains, water will leak past the valve.

Most modern valves can be changed from high-pressure to low-pressure operation either by inserting a different fitting into the inlet nozzle or by changing a detachable inlet nozzle. Some types are suitable for high or low water pressure without any alteration.

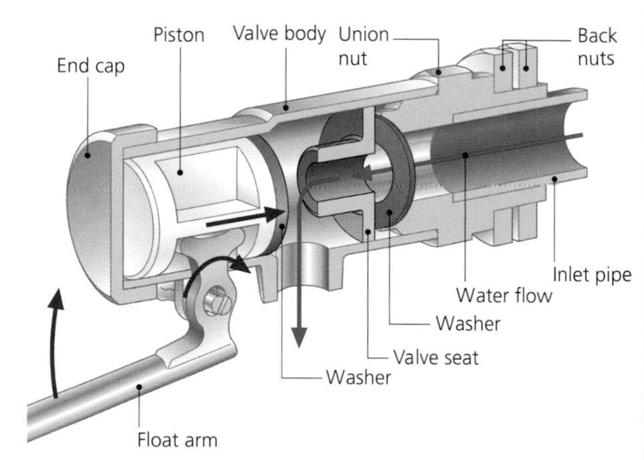

1 When the cistern is at normal level, the float (or ball) holds the arm horizontal and the valve is closed.

2 When the water level drops, the float lowers the arm and the valve opens to let more water in.

BALLVALVES IN COMMON USE

The name ballvalve is from the early type of copper ball float. Modern floats are not always balls, and ballvalves are often called float valves. Ballvalves may be made from brass, gunmetal or plastic, or may be metal with some plastic parts. The size is measured by the inlet pipe diameter; 15mm or 22mm sizes are usually needed for domestic cisterns.

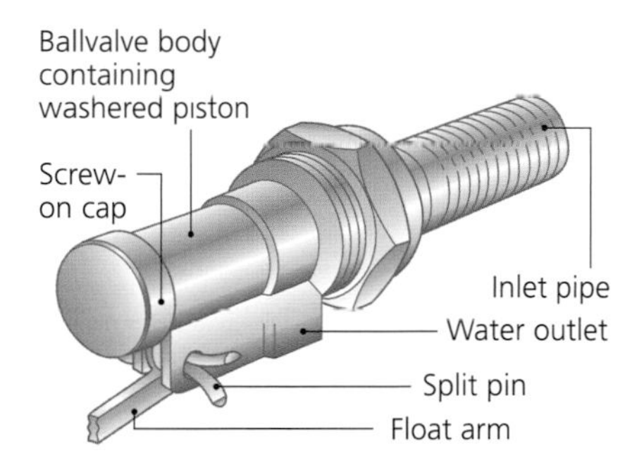

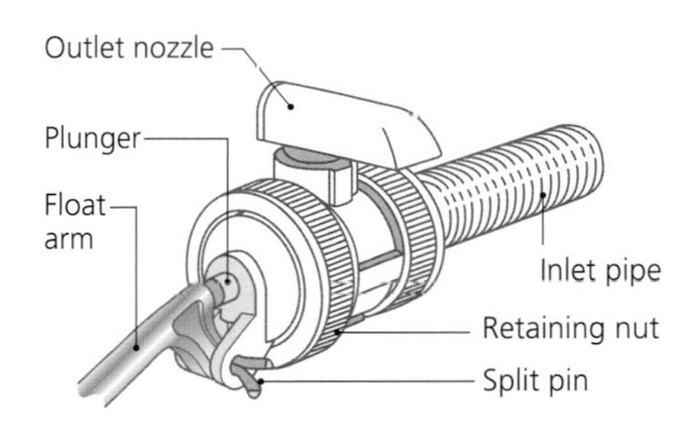

▲ **Portsmouth valve** The most common type in British homes, although no longer allowed on WC cisterns unless a double-check valve is fitted immediately before the valve. Known as Part 1 valves, Portsmouth valves are sturdy but they can be noisy and require regular maintenance. Water hammer – vibration of the rising main – can result from the valve bouncing in its seating. The bouncing is caused by ripples on the water surface when the cistern is almost full, making the float arm shake; sometimes also by the pressure of incoming water against the valve. Scale or corrosion can prevent the valve from operating properly. These valves have now been superseded by diaphragm valves. There must be a method of water adjustment other than bending the float arm.

▲ **Diaphragm valve** (also known as BRE, BRS or Garston). The water inlet is closed by a large rubber or synthetic diaphragm pushed against it by a plunger attached to the float arm. A detachable nylon overhead outlet nozzle discharges water in a gentle shower. Available with either a metal body (Part 2 valve) or a plastic body (Part 3 valve) and in high-pressure or low-pressure versions. The discharge spray cuts down filling noise and rippling, and the diaphragm keeps the plunger and float arm from contact with water, so they are not affected by scale and corrosion. The diaphragm can become worn, and grit can block the inlet chamber. The valve can be dismantled by hand for cleaning or replacement by undoing a large knurled retaining nut.

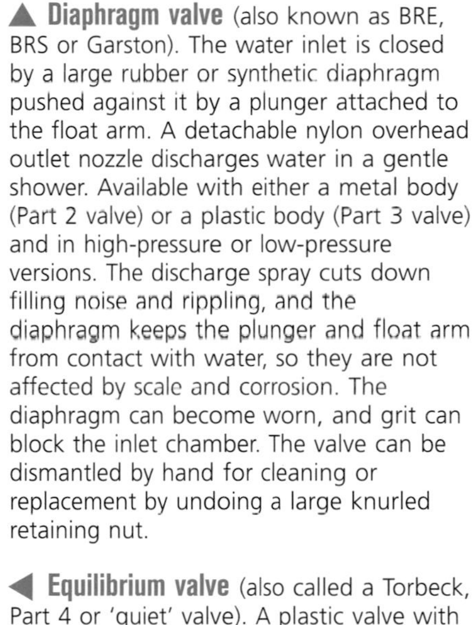

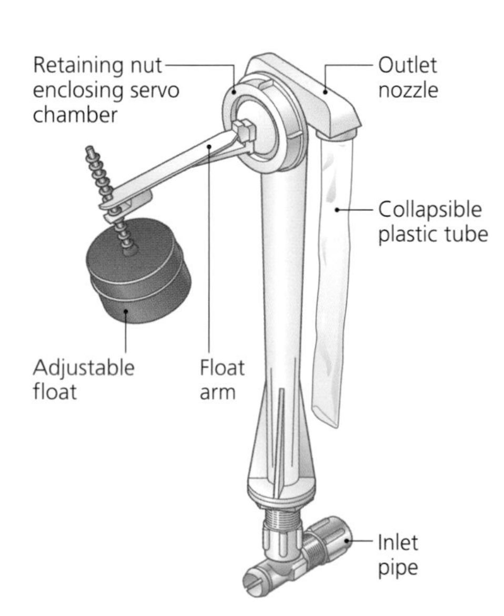

◀ **Equilibrium valve** (also called a Torbeck, Part 4 or 'quiet' valve). A plastic valve with small float and short float arm for use only in a WC cistern. Behind a diaphragm covering the inlet is a water (or servo) chamber fed via a metering pin. Equal water pressure on each side of the diaphragm keeps it closed. When the float arm drops it opens a pilot hole in the back of the chamber, covered by a sealing washer. This reduces pressure in the servo chamber, and the diaphragm opens the inlet. The outlet is overhead and via a collapsible plastic tube. Can cope with a range of water pressures: available in bottom and side entry versions. Delivery is rapid and silent. The valve can be fitted with a filter, which collects grit that might otherwise obstruct the metering pin and pilot hole. Prevents water hammer.

Adjusting the cistern water level

The normal level of a full cistern is about 25mm below the overflow outlet. The level can be raised by raising the float, or lowered by lowering the float.

Adjusting a ballvalve

Before you start If the cistern overflows, the water level is too high because the float either needs adjusting or is leaking and failing to rise to close the valve (or the valve itself may be faulty, right).

Tools *Possibly small spanner; vice.*

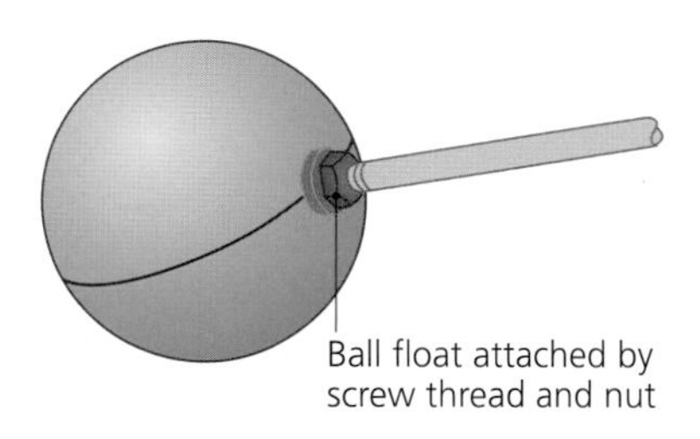

On a **Portsmouth-pattern valve** with a ball float, unscrew and remove the float from the arm. To lower the level, hold the arm firmly in both hands and bend it slightly downwards. Then refit the float. If the arm is too stiff to bend in position, remove it from the cistern and grip it in a vice.

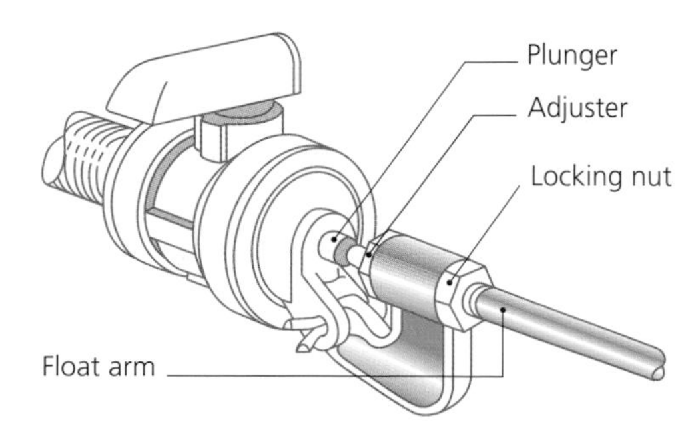

On a **diaphragm valve** with an adjuster at the top of the float arm, adjust the level by loosening the locking nut and screwing the adjuster forward, nearer to the plunger.

Alternatively Use an adjuster nut or clip near the float to move the float farther away from the valve along a horizontal arm, or to a lower position if it is linked to the arm by a vertical rod.

Repairing a faulty ballvalve

When a ballvalve does not open or close fully, it can cause airlocks and supply problems.

Portsmouth valves

Before you start If a ballvalve does not open fully, the cistern (WC or cold water storage) will be slow to fill and airlocks will occur. If the ballvalve does not close fully, the water level in the cistern becomes too high and causes a constant flow from the overflow pipe.

The water inlet of a Portsmouth valve is opened and closed by a washered piston that moves horizontally. The piston is slotted onto a float arm, which is secured with a split pin. Some types have a screw-on cap at the end of the piston. The water outlet is on the bottom of the valve in front of the float arm. The detachable inlet nozzle can be changed to suit the water pressure.

The valve will not work efficiently if the washer is worn or moving parts are clogged by limescale or corrosion.

Tools *Combination pliers; small screwdriver; fine abrasive paper; pencil. Possibly also penknife.*

Materials *Split pin (cotter pin); washer; petroleum jelly. Possibly also penetrating oil.*

1 Turn off the main stoptap if you are working on the cold water cistern, or close the gate valve or service valve to the WC cistern.

2 Use a pipe wrench to loosen the ballvalve end cap, if there is one, then unscrew and remove it.

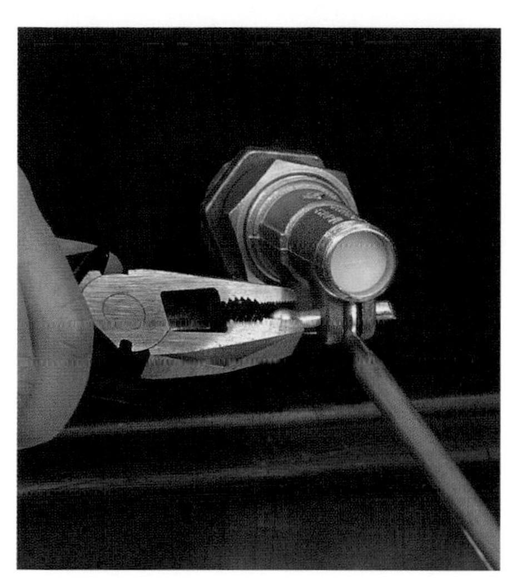

3 Use pliers to close the end of the split pin securing the float arm, then withdraw the pin. Remove the float arm and put it to one side.

4 Insert a small screwdriver blade into the slot where the float arm was seated. Use it to push the piston from the end of the casing. Catch the piston with your other hand as it comes out.

5 Clean the outside of a metal piston (but not a plastic one) with fine abrasive paper. Wrap fine abrasive paper round a pencil shaft and clean the inside of the metal valve casing.

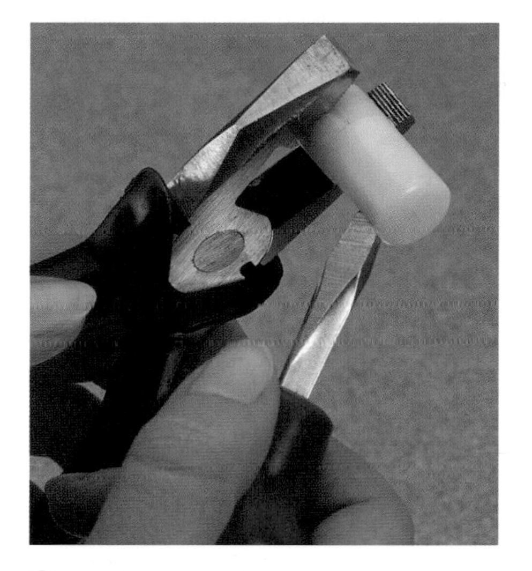

6 To renew the washer, hold the piston with a screwdriver thrust into the slot and use pliers to unscrew its washer-retaining cap. Do not force it or you may damage the piston. If a metal cap is difficult to undo, smear penetrating oil round the cap edge and try again after about 10 minutes.

7 With the cap removed, use a screwdriver to prise out the washer from the inside. (If you were unable to remove the cap, try to pick out the old washer with a penknife through the cap's open centre.)

8 Fit the new washer and screw the cap back on. Tighten with pliers. (If the cap is still on, try to force the new washer through the centre hole and push it flat with your finger.) Before refitting the piston, turn on the water supply briefly to flush out dirt from the valve casing still attached to the cistern. Lightly smear the piston with petroleum jelly before reassembling the valve and float arm. Use a new split pin to secure the arm. Restore the water supply.

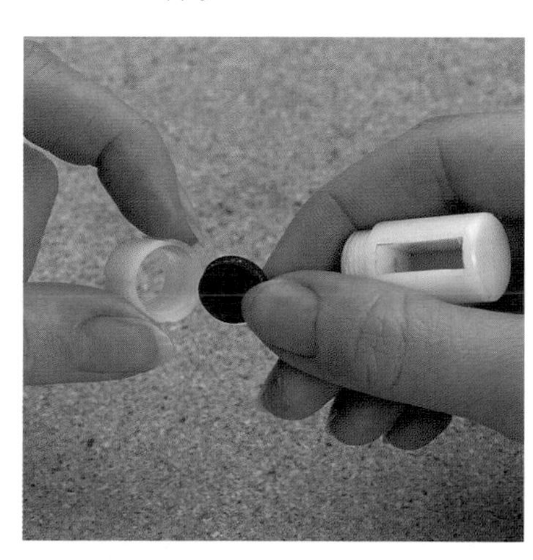

Diaphragm-type valves

The water inlet of a diaphragm-type valve is closed by a large rubber or synthetic diaphragm pushed against it by a plunger attached to the float arm. The cistern will be slow to refill if the inlet gets clogged or the diaphragm gets jammed against it.

Tools *Screwdriver. Possibly also pipe wrench; cloth for padding wrench jaws.*

Materials *Lint free rag; warm soapy water in container; clear water for rinsing. Possibly also replacement diaphragm.*

1 Turn off the main stoptap if you are working on the cold water cistern, or the gate valve or service valve on the pipe to the WC cistern (page 29).

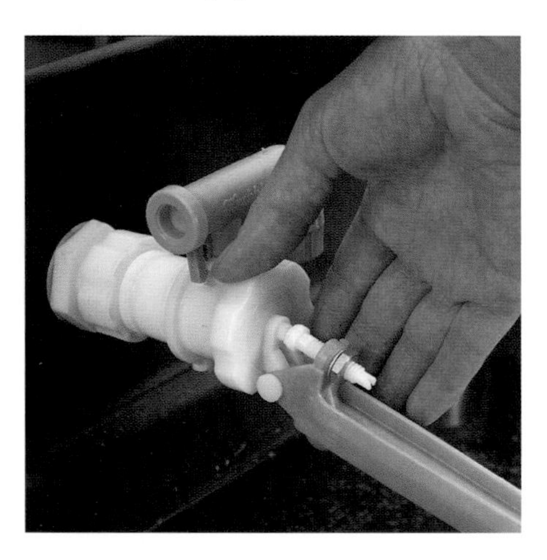

2 Unscrew the large knurled retaining nut by hand. If it is stiff, use a padded pipe wrench.

3 With the nut removed, the end of the float arm and plunger will come away. Put them to one side.

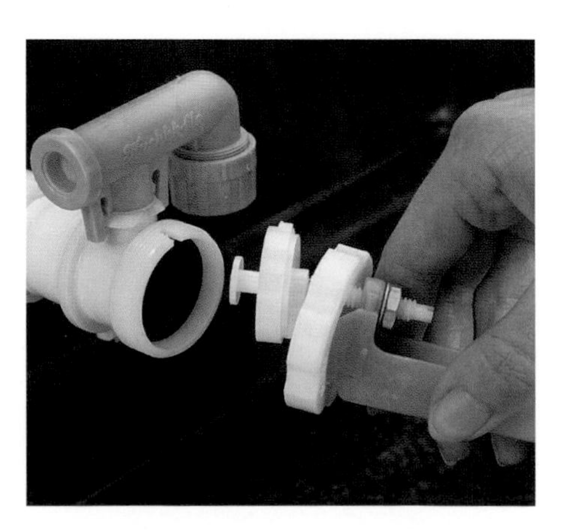

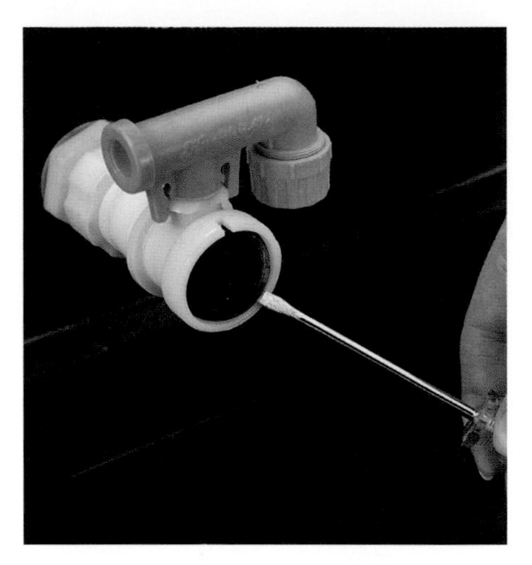

4 Use a screwdriver blade to free the diaphragm from the inlet pipe, taking care not to damage it. Note which way round it is fitted.

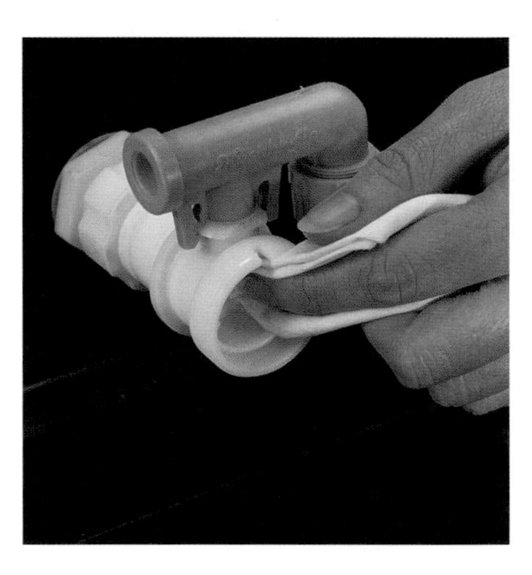

5 Use a piece of clean, lint-free rag to clean out any dirt and debris from the valve.

6 Wash the diaphragm in warm soapy water, then rinse it. If it is pitted or damaged, replace it. Fit with the rim inwards.

Alternatively If the valve is a servo type with a filter fitted, remove the filter and wash it in warm soapy water, then rinse it. If the servo valve has no filter, flush the part attached to the float arm under the tap.

7 Before reassembling the valve, turn on the water supply briefly to flush dirt or debris out of the casing. Then refit the parts and restore the water supply.

Components of a diaphragm-type valve

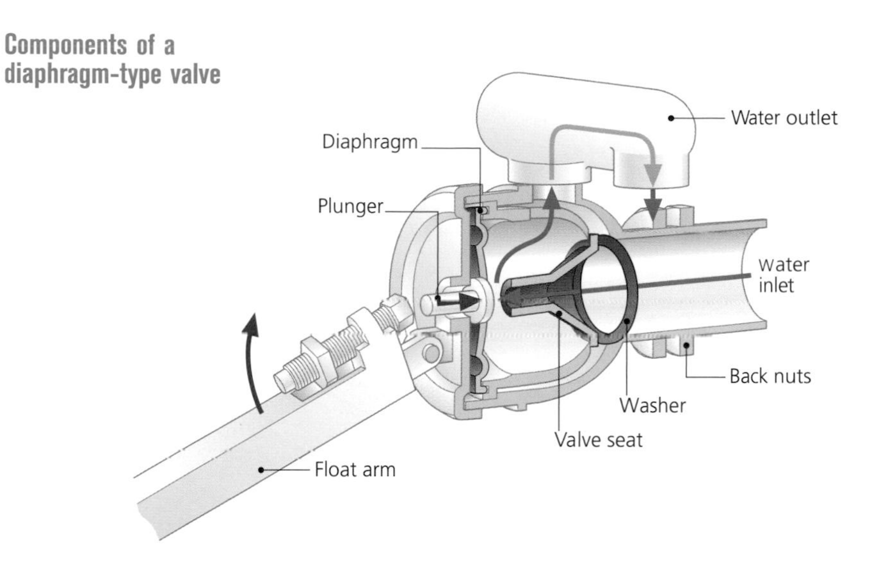

REPAIRING A LEAKING FLOAT

For a permanent repair, a new ball float must be fitted. But to get the valve back in action again until a new float is obtained, the old one can be temporarily repaired.

Tools *Small spanner; sharp knife or old-fashioned bladed tin opener; piece of wood to go across cistern.*

Materials *Plastic bag; string.*

1 Raise the float arm to close the valve and cut off the flow of the water. Then tie the arm to a length of wood laid across the top of the cistern.

2 Unscrew and remove the ball float from the float arm.

3 Find the hole through which the water is leaking and enlarge it with either a sharp knife or tin opener.

4 Drain the water from the float, then screw it back in position on the float arm.

5 Slip the plastic bag over the float and tie it securely to the float arm with string.

6 Release the arm and lower it into position to refill the cistern.

GETTING RID OF WATER HAMMER

Banging or humming from the rising main is due to the pipe vibrating when the cistern ballvalve bounces on its seating. This occurs when the float is shaken by ripples on the water as the cistern fills. Cut down the bouncing by fitting the float with a stabiliser made from a plastic pot. Hang the pot on a loop of galvanised wire from the float arm. It should trail underwater just below the float.

Ensure that the rising main is securely clipped to the roof timbers near its entry to the cistern. Also, check that a metal bracing plate is fitted on a plastic cistern to reduce distortion of the cistern wall.

The surest way to reduce water hammer is to replace a Portsmouth valve with an equilibrium valve – a type less affected by water pressure.

Fitting a new ballvalve

Fit a new ballvalve if the old one gets damaged or broken, or if you decide to change the type of valve to get rid of noise and vibration.

Tools *Two adjustable spanners.*

Materials *Ballvalve with float (of same size as existing valve); service valve (page 29) with compression fitting inlet and tap connector outlet.*

1 Turn off the main stoptap to cut off the cistern water supply.

2 Use a spanner to undo the tap connector securing the supply pipe to the valve tail. You may need to hold the valve body or any securing nut inside the cistern steady with a second spanner.

3 Disconnect the supply pipe.

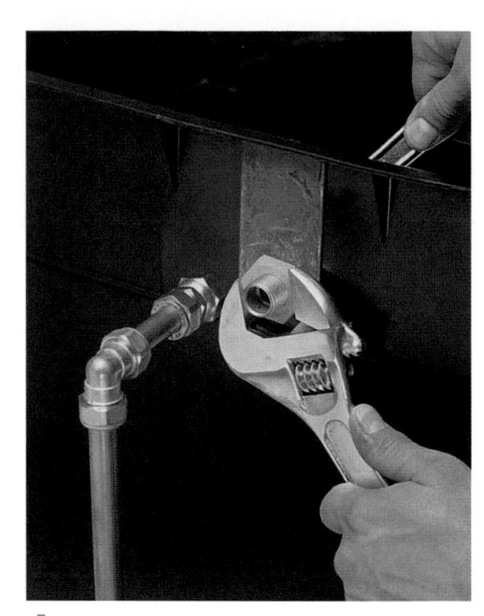

4 Undo the backnut securing the ballvalve to the cistern. Remove the valve.

5 Take off the backnut from the new ballvalve and put it aside.

Ballvalve fixing to cistern The valve is held against the wall of the cistern by two backnuts on its threaded inlet tail.

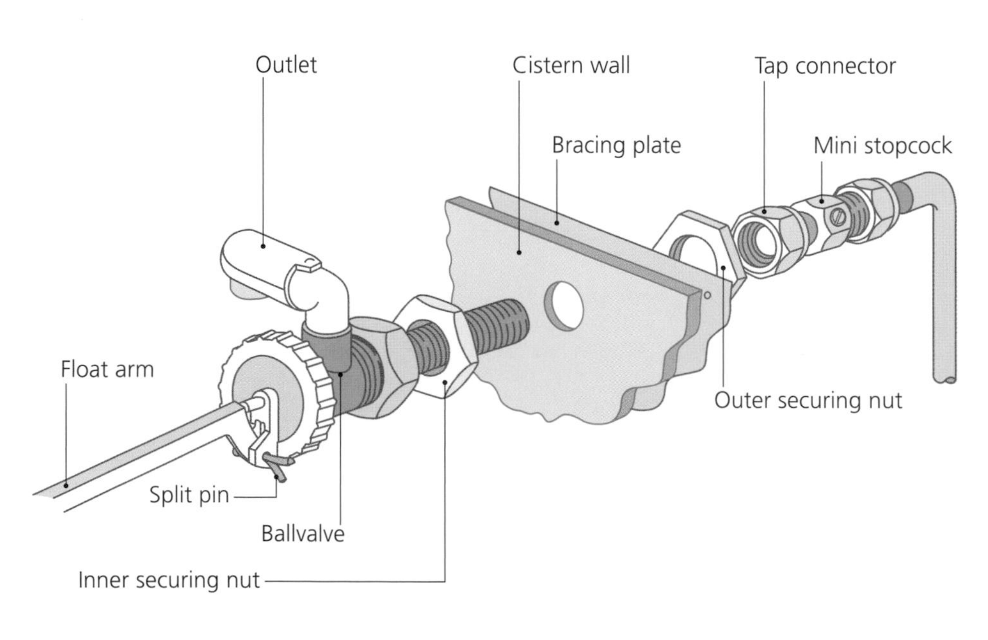

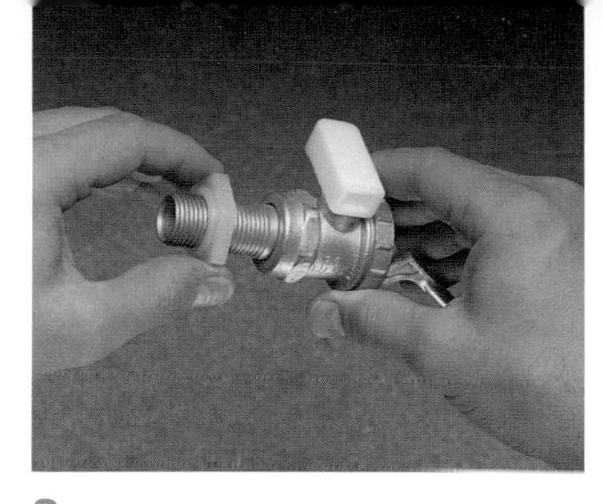

6 Slip the inner securing nut over the new valve tail and push the tail through the cistern and the bracing plate from the inside.

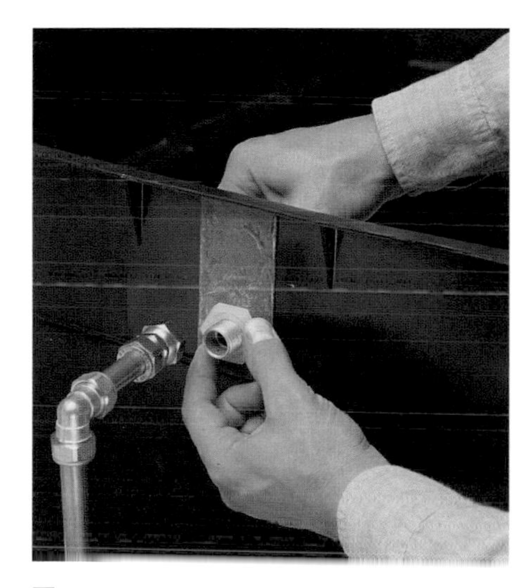

7 Screw on the outer backnut by hand. Tighten it by half a turn with a spanner.

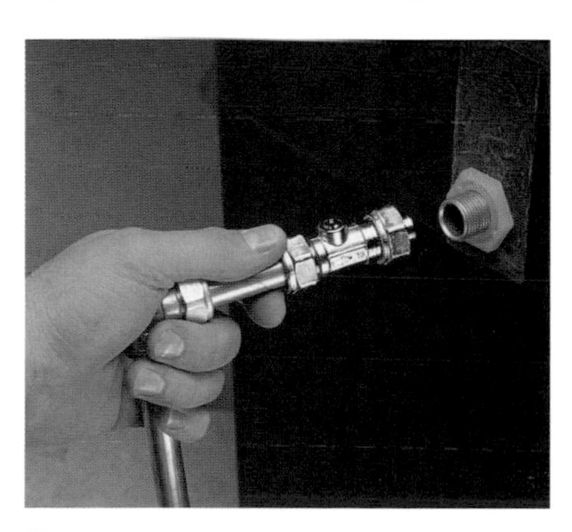

8 Remove the existing tap connector and fit the service valve. Screw the connector nut to the new valve tail.

9 Restore the water supply, making sure the service valve is fully open. Adjust the cistern water level (page 48).

Clearing a blocked WC

The usual faults with a WC pan are blockages or leaks. A leak from the pan outlet is not difficult to repair, but a cracked pan will have to be replaced.

Clearing a washdown pan

When a washdown bowl is flushed, the two streams of water, one from each side of the rim, should flow equally to meet at the front. The water should leave the pan smoothly, not eddying like a whirlpool. If the cistern is working properly but the bowl fails to clear, something is obstructing either the flush inlet or the pan outlet.

If the flush water rises almost to the pan rim, then ebbs away very slowly, there is probably a blockage in the pan outlet (or possibly in the soil stack or drain into which it discharges).

Tools *WC plunger. Possibly also flexible drain auger; bucket; mirror; a pair of rubber gloves.*

1 To clear the pan, take the plunger and push it sharply onto the bottom of the pan to cover the outlet. Then pump the handle up and down two or three times.

2 If this does not clear the pan, use a flexible drain auger to probe the outlet and trap.

3 If the blockage persists, check and clear the underground drain.

4 Flush the cistern to check that water is entering the pan properly, with streams from each side of the rim flowing equally to meet at the front.

5 If the flow into the pan is poor or uneven, use a mirror to examine the flushing rim. Probe the rim with your fingers for flakes of limescale or debris from the cistern that may be obstructing the flush water.

Alternatively If you have no WC plunger, you may be able to use a mop. Or stand on a stool and, all in one go, tip in a bucket of water.

Clearing a siphonic pan

Blockages are more common in siphonic pans because of the double trap and the delicate pressure reducing pipe seal (also known as the atomiser seal). Do not use a plunger on a blocked siphonic pan because this can dislodge the seal.

A blockage can usually be cleared with an auger or by pouring several buckets of warm water into the pan. But if, after clearing the blockage, the water still rises in the pan as it is flushed, renew the seal.

Tools *Screwdriver; adjustable spanner; container for bailing or a tube for siphoning; silicone grease.*

Materials *Pressure reducing pipe seal.*

1 Remove the cistern (page 45) and locate the pipe protruding from the bottom of the siphon.

2 Remove the rubber mushroom-shaped seal and fit a replacement. Lubricate the new seal with silicone grease so that it will slide down the pipe.

3 Refit the cistern and test the flush. The water should be removed from the bowl with a sucking noise before the clean water comes in from the rim of the bowl.

Repairing a leaking pan outlet

A putty joint may leak when the putty gets old and cracked.

To replace a putty joint with a push-fit connector (page 78), the pan must be moved forward then refitted. Alternatively, you can repair the joint using waterproof building tape or non-setting mastic filler.

Chip and rake out the old putty with an old chisel and bind two or three turns of tape round the pan outlet. Then poke more tape firmly into the rim of the soil-pipe inlet. Fill the space between the rim and pan outlet with mastic. Bind two more turns of tape round the joint.

PLUMBING AND EARTHING ▶

All metal pipework must be joined to the house's main earth bonding system so that it is impossible for someone touching exposed metalwork to be electrocuted if it becomes live.

Plastic pipe and fittings are non-conducting so if either has been used in a run of metal pipe, a 'bridging' earth wire must be fitted between the metal sections. When installing a metal bath or sink, follow the instructions on page 88.

Choosing pipes and fittings

Copper and plastic are the two most common materials used in household pipework. Stainless steel piping can be used in the same way as copper: although it is more expensive and harder to work with, it can be used safely with galvanised steel radiators, without the corrosive reaction that copper produces. Lead and iron pipes are no longer used, although they may still be found in older buildings.

Pipe sizes Domestic water supply pipes are made in metric sizes. Copper pipe is measured by the outside diameter. Standard sizes are 15mm, 22mm and 28mm. Older supply pipes were made in imperial sizes and were measured by the inside diameter. Imperial and metric piping can be joined, but an adaptor is needed for some sizes, depending on the type of joint and connector used.

Plastic supply pipes are made in the same nominal sizes as copper pipes, but have thicker walls. Medium-density polyethylene (MDPE) pipe is now widely used for underground supply pipes, and is coloured blue. Semi-flexible polybutylene (PB) and cross-linked polyethylene (PEX) pipes can both be used for indoor hot and cold supply pipes. They come in white or grey, and long lengths are available.

Types of pipe

Copper

For hot and cold supply pipework indoors. Sold in three sizes (15, 22 and 28mm) in 2m and 3m lengths. Join with compression joints, with soldered capillary joints or with push-fit joints. Cut with a hacksaw or pipe cutter.
Advantages Withstands high temperatures. Bends easily round corners, and economical to use. Can be painted. Rigid, so needs few supports: 15mm pipe every 1.2m horizontally, 1.8m vertically; 22mm and 28mm pipe every 1.8m horizontally, 2.4m vertically.
Disadvantages Pipes hot to touch. May split or burst if water freezes. Brings about the corrosion of galvanised (zinc-coated) steel if joined directly to it.

Semi-flexible plastic For hot and cold pipework indoors. Sold in 15mm, 22mm and 28mm sizes in straight 2m and 3m lengths and in rolls up to 100m long. Join with plastic push-fit joints or brass compression joints (always with a metal supporting insert). Cut with a trimming knife or with special pipe cutters shaped like secateurs.
Advantages Easy to cut and to bend in gentle curves or round corners (metal bend supports available) and can be run through holes cut in floor joists. Joints easy to make (and re-make) and allow pipe to be twisted in fitting (important for tee joints). Long lengths minimise number of fittings needed. Can be painted with water-based or solvent-based (but not cellulose) paints. Widely available with full range of fittings. Insulates well – pipes not too hot to touch. Pipes less likely to freeze in cold weather but cannot withstand too high temperatures – so must not be used close (within 1m) to boiler for central heating pipe runs. Does not corrode or support scale formation.
Disadvantages Needs more support than copper to prevent sagging: 15mm pipe every 0.3m horizontally and 0.5m vertically; 22mm pipe every 0.5m horizontally and 0.8m vertically. Pipe cheaper than copper; fittings cost about the same as compression fittings for copper (though fewer needed). Extra earthing needed if replacing length of copper pipe (see opposite).

BRASS IN ACID WATER ▶
In areas where the water supply is particularly acid, check with the regional water company before using brass joints (see next page) – especially if you are using copper piping. Brass is an alloy of zinc and copper, and in highly acid water a reaction between the two metals can cause the zinc to dissolve and the joint to fail. This process is known as dezincification. Use joints made of gunmetal or DR (dezincification-resistant) metal instead.

CHOOSING THE TYPE OF JOINT

There are several different joint types available. Some can be used for more than one pipe type. All the types are available as straight couplers, elbows and tees for joining pipework, and as threaded joints for connecting pipes to taps, ballvalves, storage cisterns and hot water cylinders.

Compression joint

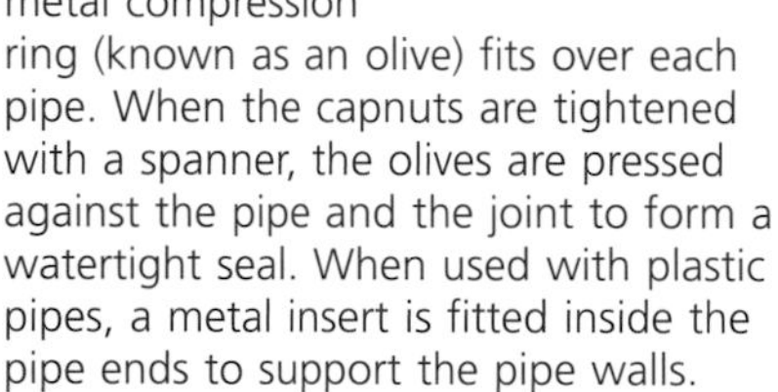

A brass joint with screw-on capnuts at each end. A soft metal compression ring (known as an olive) fits over each pipe. When the capnuts are tightened with a spanner, the olives are pressed against the pipe and the joint to form a watertight seal. When used with plastic pipes, a metal insert is fitted inside the pipe ends to support the pipe walls.
Advantages Widely available. Can be re-used if dismantled (with a new olive).
Disadvantages More expensive than a capillary joint. Looks clumsy. Tightening can be difficult in awkward places.

Capillary joint

A copper joint with internal pipe stops, inside which copper pipe ends are sealed with solder heated by a blowtorch.
There are two types:

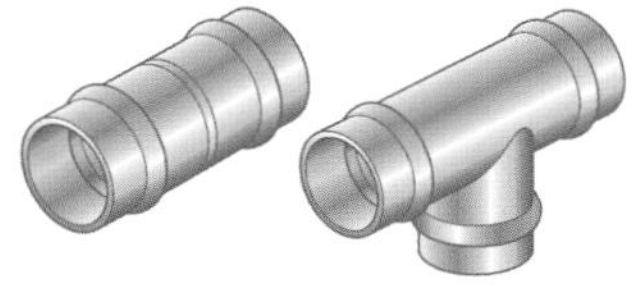

Soldered integral-ring type

Solder-ring joint Has a built-in ring of solid solder near each end. This is melted and flows round the pipe by capillary action as the joint is heated. Also known as a integral-ring, pre-soldered or Yorkshire joint.
End-feed joint Contains no solder. Wire solder is fed into each end of the fitting as the joint is heated, and is drawn in by capillary action. Cheaper than solder-ring joints.

End-feed joint

Advantages Widely available. Cheaper than all the other joints. Neat finish.
Disadvantages Only for copper pipe. Cannot be re-used if dismantled. Tricky to use successfully. Special joints needed to link new metric and old imperial pipe sizes.

Plastic push-fit collet joint

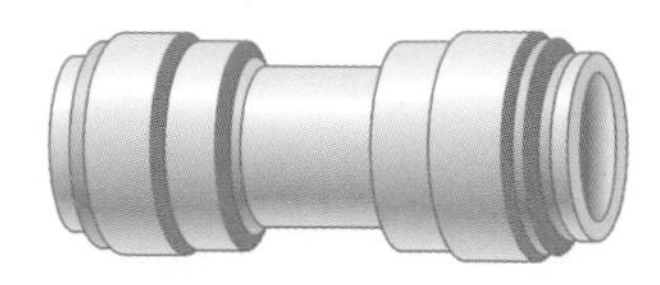

A plastic joint with a toothed ring (called a collet) to grip the pipe and an O-ring seal. The pipe end is pushed into the joint, and can be released by holding the ring against the joint as the pipe is withdrawn. Suitable for both copper and plastic pipe.
Advantages Quick and easy to use. Extremely reliable. Can be dismantled easily and re-used. Pipe can be rotated in joint for branch alignment. Fitting unobtrusive in appearance.
Disadvantages Not suitable for connection directly to boiler. More expensive than capillary fittings.

Plastic grab-ring push-fit joint

A plastic joint with an internal grab ring and O-ring seal. The pipe end is pushed into the joint, and can be released by undoing the cap nuts and prising open the grab ring with a special tool. Suitable for both copper and plastic pipe.
Advantages Very easy to use. Can be undone and re-used if new grab ring is fitted. Pipe can be rotated in joint for branch alignment.
Disadvantages More expensive than compression fittings. Bulky.

Metal push-fit

A sleek copper or brass fitting with an internal seal. Can be released with special tool. Suitable for use with copper and plastic pipe (with pipe insert).

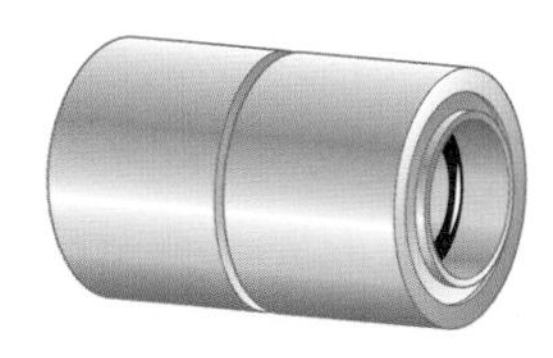

Advantages No tools needed to make fitting. Maintains earth continuity when used with copper. Neat in appearance.
Disadvantage Fairly expensive.

Threaded joints

Joints for connecting pipes to taps, cylinders and cisterns have a screw thread at one end which may be internal (female) or external (male). Threaded joints must be matched male to female. Made watertight by binding PTFE thread-sealing tape round the male thread.

Ways of joining pipes

Preparing the pipe ends

Before two pipe lengths of any material can be joined, the ends must be cut square and left smooth. Copper pipe needs careful cutting and finishing to ensure watertight joints. You can cut plastic pipes with special cutters or with a sharp craft knife.

Tools *Pipe cutter or hacksaw; half-round file. Possibly also vice or portable workbench.*

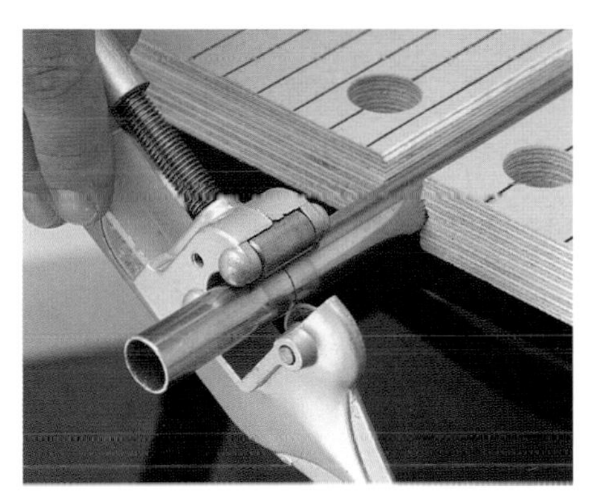

1 Cut the pipe ends square using a pipe cutter or hacksaw (not on plastic pipe). Holding the pipe in a vice while sawing helps to ensure a square cut.

2 Smooth away burrs inside the cut ends of copper pipe with the reamer on the pipe cutter. Use a file to smooth the end and the outside of the pipe.

Making a compression joint

This is a strong and easy method of joining copper and plastic pipes. Tightening the nuts correctly is critical – the joint will leak if they are not tight enough or if they are over-tightened.

Tools *Two adjustable spanners (with jaw openings up to 38mm wide for fittings on 28mm piping); or, if you have any that fit, two open-ended spanners – capnut sizes on different makes of fittings vary.*

Materials *Compression fitting.*

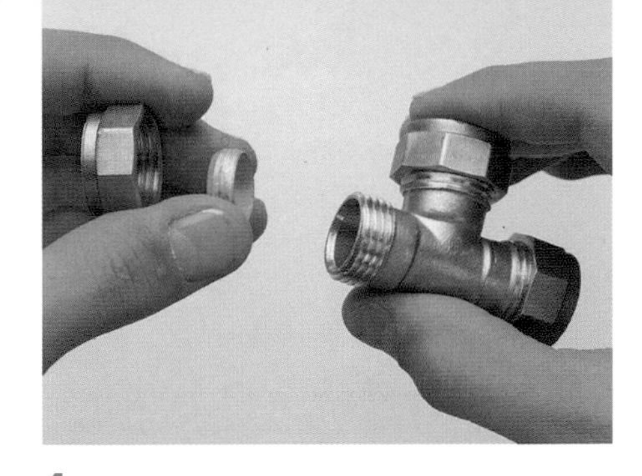

1 Unscrew and remove one cap nut from the fitting. If the olive has two sloping faces rather than a convex one, note which way round it is fitted as you remove it.

2 Take one pipe and slide the cap nut over it, then the olive. Make sure the olive is the correct way round if it has two sloping faces. Fit a pipe support sleeve to plastic pipe.

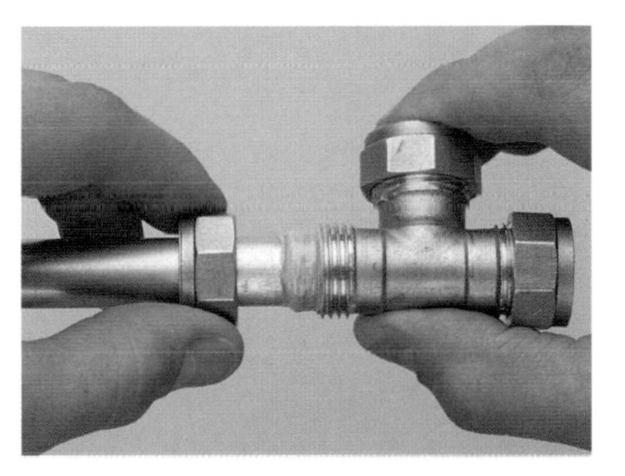

3 Push the pipe into one end of the fitting up to the internal pipe stop. Then slide the olive and nut up to the fitting and hand-tighten the cap nut.

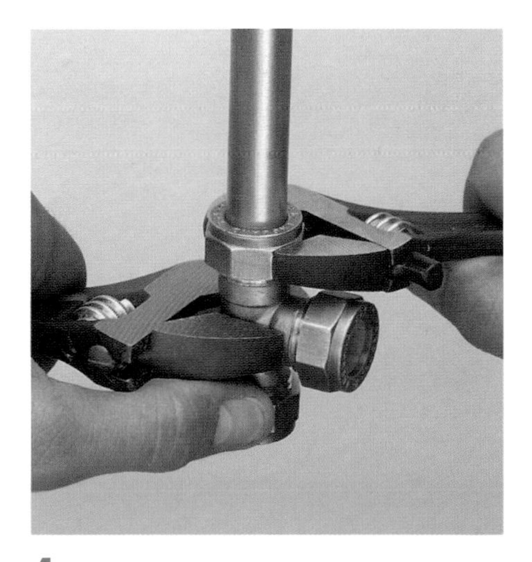

4 Hold the body of the fitting securely with one spanner while you give the cap nut one and a quarter turns with the other. Do not overtighten it further. Fit pipes into other openings of the fitting in the same way.

Making a soldered joint

This is a more difficult joint for the amateur plumber to make successfully. Too little heat will fail to make a complete solder seal inside the fitting, while too much heat will make all the solder run out.

Tools *Wire wool or fine emery paper; blowtorch; clean rag.*

Materials *Tin of flux; soft lead-free solder wire for end feed joints only. Possibly also sheet of glass fibre or other fire-proof material for placing between the joint and any nearby flammable material.*

1 Clean the ends of the pipes and the inside of the fitting thoroughly with wire wool or fine emery paper. They must show clean, bright metal to make a successful joint.

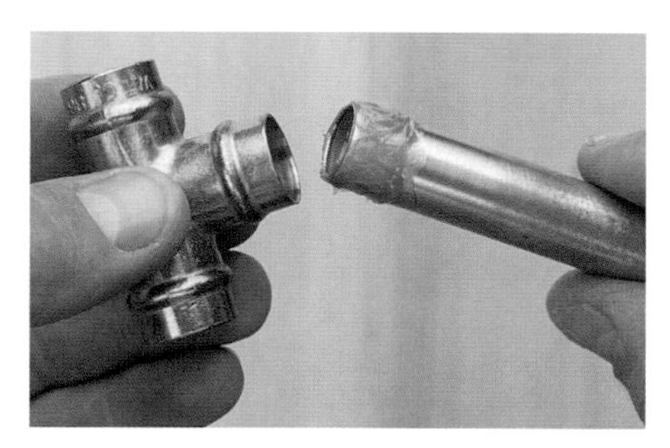

2 Smear the cleaned surfaces with flux, which will ensure a clean bond with the solder. Push the pipe into the fitting as far as the pipe stop. For an integral-ring fitting, push pipes into all the openings because all the solder rings will melt once heat is applied to the fitting. Wipe off excess flux with a clean rag, otherwise the solder will spread along the pipe surface.

3 Fix the pipe run securely in position with pipe clips and position fire-proof mat.

4 For an integral-ring fitting, heat the joint with a blowtorch until a silver ring of solder appears all round the mouths of the joints. Solder all the joints on the fitting in the same operation.

Alternatively For an end feed fitting, heat the joint until you see flux vapour escaping. Then remove the heat (otherwise the solder will melt too fast and drip) and apply soft solder wire round the mouth of the fitting and heat again until a silver ring of solder appears all round. If you have to leave some joints of the fitting until later, wrap a damp cloth round those already made to stop the solder re-melting.

5 Leave the joint undisturbed for about five minutes while it cools.

Making a push-fit joint

This is a simple method that can be used to join both copper and plastic pipes. The only tools needed are those used to cut and smooth the pipe ends. Both plastic and metal push-fit fittings can be used on either copper or plastic pipe – with plastic pipe, pipe inserts must be used. You will also need adaptors for joining to old-size (imperial) copper pipe.

Tools *Pencil; measuring tape, pipe cutters (or hacksaw for copper pipe), trimming knife (for plastic pipe).*

Materials *Push-fit joint.*

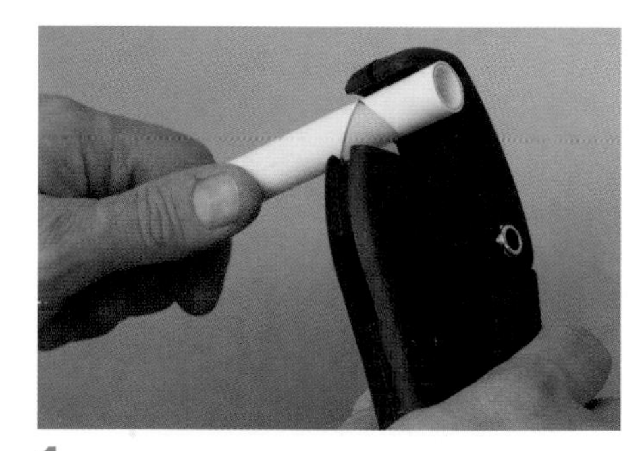

1 Cut plastic pipe using the special pipe cutters (secateurs) available – some plastic pipe is marked with insertion distances and you cut it at an insertion mark. Cut copper pipe as shown on page 57.

2 Check that the cut pipe end is smooth, otherwise sharp edges coud damage the O-ring seal in the fitting and cause the joint to leak. Use a file on copper pipe and a trimming knife on plastic pipe.

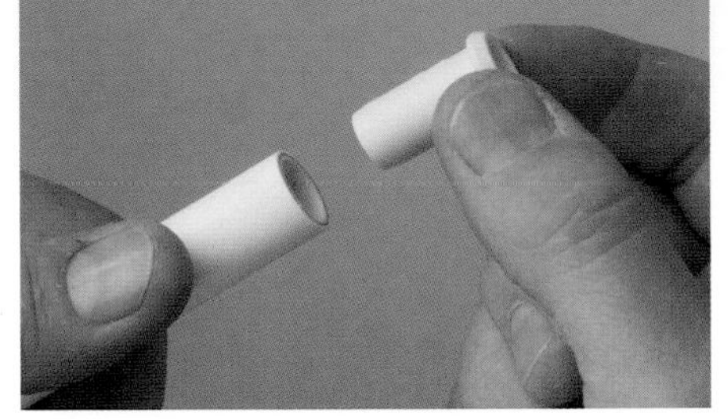

3 For plastic pipe, push the support sleeve into the pipe end with a twisting motion.

4 Unless you have cut plastic pipe at an insertion mark, use a pencil to make a mark on the pipe for the correct insertion distance (check instructions supplied with the fitting).

5 Push the pipe into the fitting up to the insertion mark. If it is not pushed fully home the pipe will blow out under pressure. Check that the pipe is properly home by tugging gently on it. Fit a bridging earth wire (see page 55) if you have used a plastic push-fit fitting in a run of copper pipework.

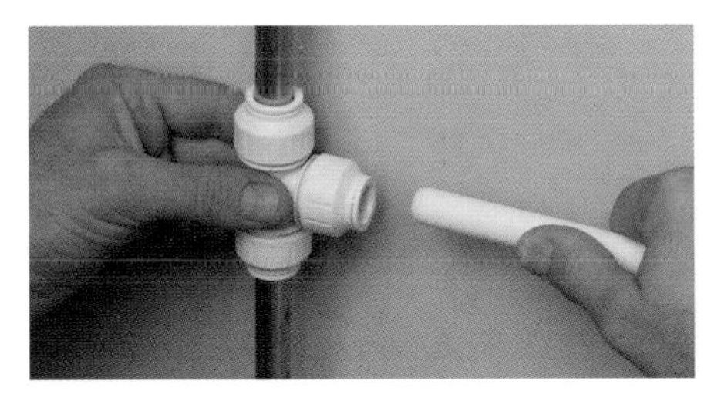

Reusing fitting Check which method is needed to de-mount a push-fit fitting so it can be remade. With some slimline fittings, you just push in the collet; with others you need some kind of de-mounting tool or collar. With grab-ring fittings, you unscrew the large retaining cap, lever out the old grab ring and fit a new one. Take care – it's sharp.

Bending copper pipe

Never try to bend rigid copper pipe by hand without a spring to support the pipe walls – the pipe will kink at the bend if it is not supported.

Tools *Bending springs of the required diameter (15mm or 22mm); or pipe-bending machine with pipe formers and guide blocks; screwdriver; length of string.*

Materials *Petroleum jelly.*

Bending with a spring

1 If the pipe is longer than the spring, tie string to the spring end.

2 Grease the spring well with petroleum jelly and push it into the pipe.

3 Bend the pipe across your knee with gentle hand pressure to the required angle.

4 Overbend the pipe a little more, then ease it gently back again. This action helps to free the spring and makes it easier to withdraw.

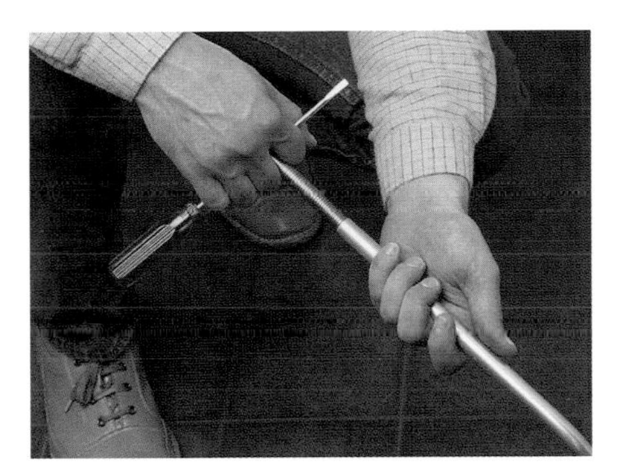

5 Insert a screwdriver blade through the loop at the end of the spring. Twist the spring to reduce its diameter, then pull it out.

Bending with a machine

1 Clamp the pipe against the correct-sized semicircular former.

2 Place the guide block of the correct diameter between the pipe and the movable handle.

3 Squeeze the handles together until the pipe is curved to the required angle round the semicircular former.

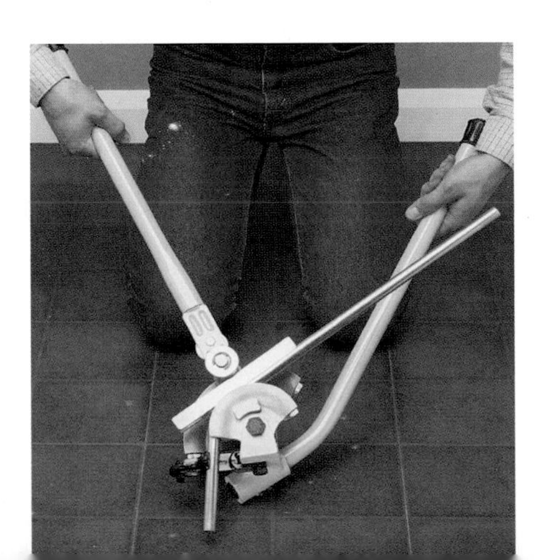

Choosing waste pipes and traps

All waste water outlets are fitted with a trap – a bend in the piping that retains water and stops foul air from the drain getting back into the room.

Plastic waste pipes are made in 40mm and 50mm diameter for sinks, baths and shower trays, 32mm for washbasins and 22mm for overflow pipes from cisterns. Pipes and fittings from different makers are not always interchangeable.

PVCu For cold water overflow pipes from WC and storage cisterns. Joined by push-fit or ring-seal joints or by solvent welding. White, grey, brown, sometimes black. Sold in 3m and 4m lengths.

PVCmu (modified un-plasticised polyvinyl chloride) For hot waste from sinks, baths, washbasins and washing machines. Joined by push-fit or ring-seal joints, or by solvent welding (see page 63). Ring-seals for connection to main stack. Grey, white, sometimes black. Sold in 4m lengths.

Polypropylene For cold-water overflow pipes and hot waste from sinks, baths, washbasins and washing machines. Joined by push-fit or ring-seal joints only. White and black. Has slightly waxy surface. Sold in 3m and 4m lengths.

FITTINGS FOR WASTE PIPES

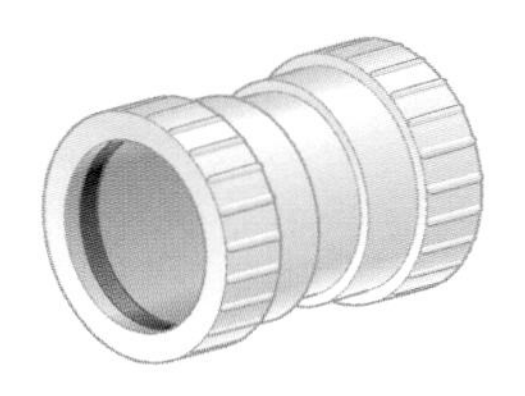

Locking push-fit (ring-seal) joint
Polypropylene joint with screw-down retaining cap nuts. The sealing ring is usually ready-fitted.

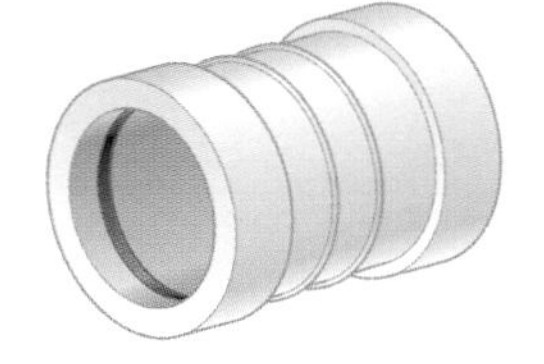

Push-fit (ring-seal) joint
Rigid polypropylene sleeve with a push-fit connection. Cannot be connected to any existing copper, steel or plastic system unless a locking ring fitting is interposed.

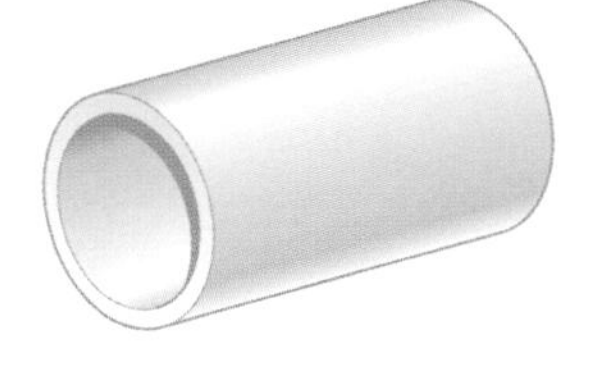

Solvent-weld joint
A plastic sleeve with a built-in pipe stop at each end. Used with PVC piping, which is secured in the sleeve by means of a strong adhesive, recommended by the joint manufacturer.

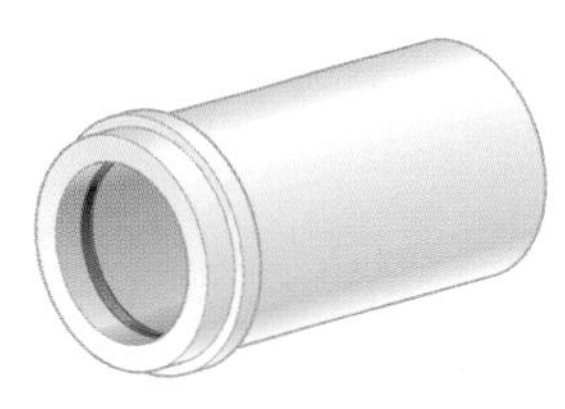

Expansion coupling
PVCmu joint designed for a solvent-weld joint at one end and a ring-seal joint at the other. Because solvent-weld joints do not allow for heat expansion, the coupling should be inserted every 1.8m in a long run of solvent-welded waste pipe.

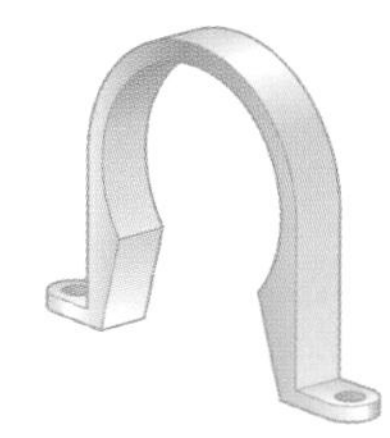

Pipe strap
Straps or clips for supporting waste pipes are available in compatible sizes. For sloping pipes they should be fixed about every 500mm. The slope must be at least 20mm per 1m run (page 71). For vertical pipes, fix clips every 1.2m.

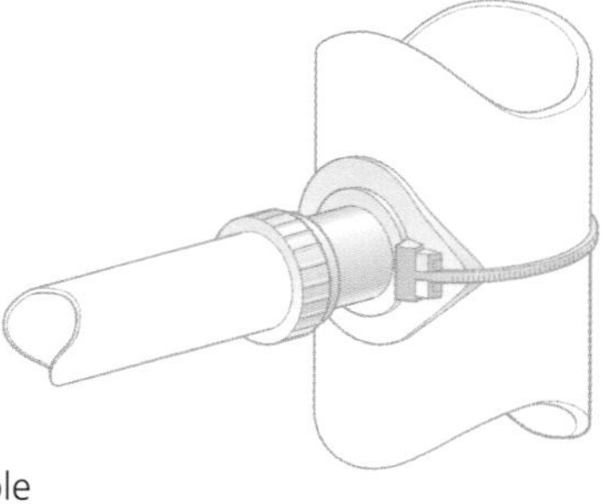

Stack connector
A clip-on polypropylene boss for fitting a new waste pipe into the stack, in which a hole has to be cut.

CHOOSING A TRAP

There are different types of trap for different situations. There may also be different outlet types (vertical or horizontal) and different seal depths. The seal – the depth of water maintained in the trap – is normally 38mm, but a trap with a deep seal of 75mm must, by law, be fitted to any appliance connected to a single-stack drainage pipe (see page 17). This guards against the seal being destroyed by an outflow of water, allowing foul air from the stack to enter the house.

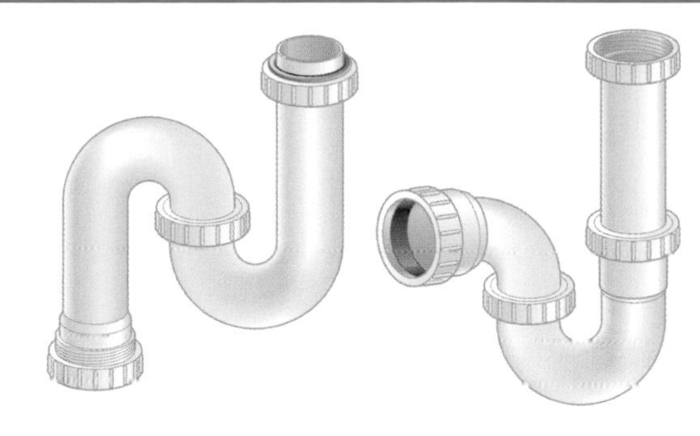

Tubular traps

A two-piece trap for a sink or basin, with an S (down-pointing) outlet (above left). Traps are also available with a P (horizontal) outlet (above right) and an adjustable inlet to allow an existing pipe to be linked to a new sink at a different height. Tubular traps are cleaned by unscrewing the part connected to the sink waste outlet.

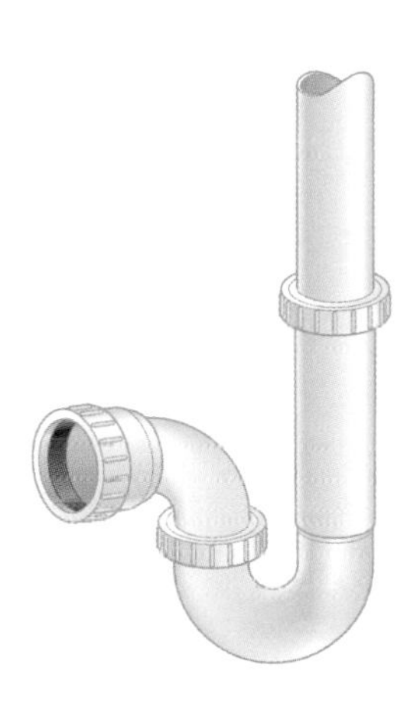

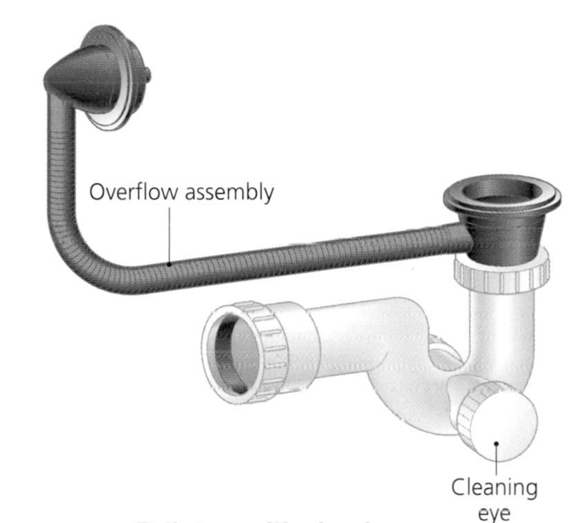

Washing-machine trap

A tubular trap with a tall stand-pipe for the washing-machine waste hose, and an outlet to link to the waste pipe.

Bath trap with cleaning eye and overflow pipe

A cleaning eye can be unscrewed to clear a blockage and is useful where access is difficult. A flexible overflow pipe can be connected to a side or rear inlet on some bath traps. The overflow is a safeguard in case a tap is left running while the bath plug is in.

Standard bottle trap

Normally used only for washbasins, which have a small outflow. Most have a P outlet, but an S converter may be available. Some have a telescopic tube to adjust to different heights.

Anti-siphon bottle trap

Designed to allow air to enter the trap and prevent the seal being lost. Use where there is an occasional heavy flow, or long, steep pipe run.

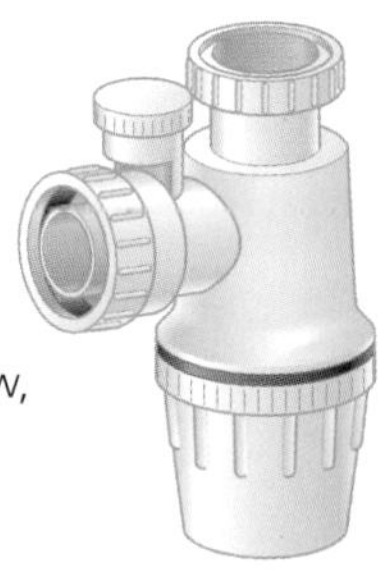

Fitting a waste pipe and trap

Before 1939, waste and overflow pipes for sinks, washbasins, baths and cisterns were made of lead or galvanised steel. After that, copper was used until about 1960. Since then plastic has been in general use.

Joints for joining waste pipes come in broadly the same configurations as those for joining supply pipes; some additional joints are shown on page 60.

If you plan to fit a new pipe that has to be connected into a soil stack (page 17), get the approval of your local authority Building Control Officer.

Making a push-fit joint

Push-fit or ring-seal joints must be used to connect polypropylene waste pipes, as these cannot be solvent-welded. Joints can be re-used with a new seal.

Tools *Hacksaw; sharp knife; clean rag; newspaper; adhesive tape; pencil.*

Materials *Push-fit joint; silicone grease.*

1 Wrap a sheet of paper round the pipe as a saw guide. Cut the pipe square with a hacksaw.

2 Use a sharp knife to remove fine shavings of polypropylene and any rough edges from the pipe.

3 Wipe dust from inside the fitting and the outside of the pipe.

4 On a locking-ring connector, loosen the locking ring.

Locking-ring connector

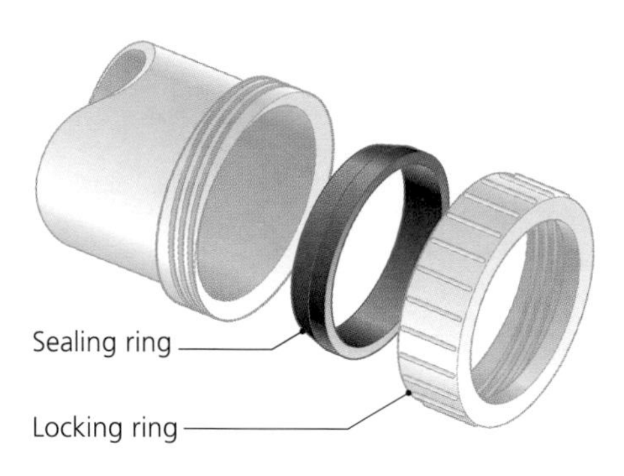

5 Make sure that the sealing ring is properly in place (above), with any taper pointing inwards. If necessary, remove the nut to check.

6 Lubricate the end of the pipe with silicone grease.

7 Push the pipe into the socket as far as the stop – a slight inner ridge about 25mm from the end. This allows a gap of about 10mm at the pipe end for heat expansion.

Alternatively If there is no stop, push the pipe in as far as it will go, mark the insertion depth with a pencil, then withdraw the pipe 10mm to leave an expansion gap.

8 Tighten the locking ring.

HELPFUL TIP

When running waste pipes along a wall, hold them in place with waste pipe clips.

Use wallplugs and screws to attach the clips to a masonry wall, or hollow-wall fixings on plasterboard.

Remember to check for pipes and cables before you drill.

Making a solvent-welded joint

Because solvent-welded joints are neat they are suitable for exposed PVCmu pipework. However, they are permanent and should be used only where they will not need to be disturbed. Push-fit connections are used at traps, where the joint may need to be undone occasionally.

Tools *Hacksaw; half-round file, cloth.*

Materials *Solvent-weld cement; appropriate connector (see page 60); appropriate pipe.*

1 Cut the pipes to the required length with a hacksaw, remove the burrs inside and out with a half-round file, and wipe thoroughly with a clean cloth.

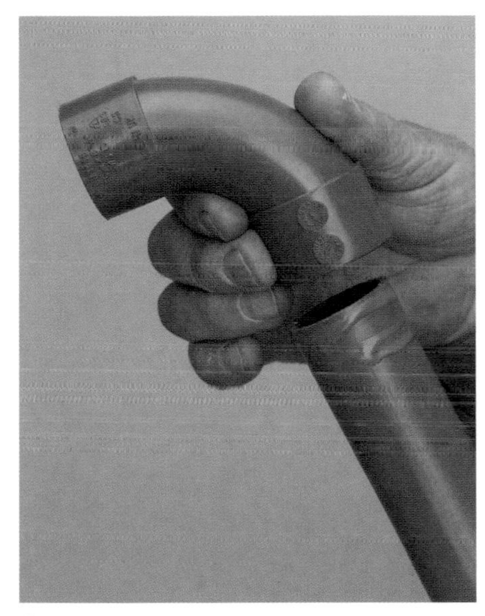

2 Apply solvent-weld cement around the end of the pipe and push it into the joint.

3 Wipe off excess cement with the cloth and allow the joint to dry before moving on to the next joint.

Fitting a trap

Traps are either tubular or bottle-shaped (see Choosing a trap, page 61), and are made in suitable sizes to fit between a sink, bath or washbasin waste outlet and its waste pipe.

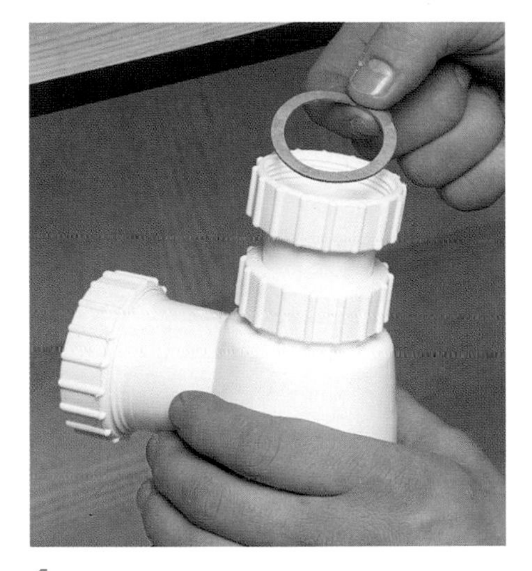

1 Check that the locking nut on the trap inlet is unscrewed and the rubber washer in position.

2 Push the trap inlet into the waste outlet and screw the nut onto the waste outlet thread.

3 Connect the trap outlet to the waste pipe with a push-fit joint.

Choosing a tap

Most taps work in the same way – turning the handle opens or closes a valve that fits into a valve seat. The valve – a rod and plate known as a jumper valve – is fitted with a washer that is replaced when it is worn and the tap drips.

CERAMIC WASHERS

The latest tap designs incorporate hardwearing ceramic discs – one fixed, one moving – instead of washers, and when the tap is turned on, openings in the discs line up so that water flows through. They open or close with only a quarter turn (through 90°) of the handle.

The discs become smoother and more watertight with wear, so should never need replacing but, in hard water areas, performance may eventually be affected by limescale.

Mixed measurements

New plumbing fittings come in metric sizes, but if the plumbing in your house dates from before the mid 1970s, your pipework is in imperial sizes (½in, ¾in and 1in inner diameter). Modern copper pipe comes in outer diameters of 10mm, 15mm, 22mm and 28mm. Choose 15mm pipe to join to existing ½in pipework and 28mm for 1in pipework. You will need special connectors for joining 22mm pipe to ¾in pipework.

Pillar tap The type still often used in bathrooms, with a vertical inlet that fits through a hole in the sanitaryware. The conventional tap has a bell-shaped cover – generally known as an easy-clean cover – and a capstan (cross-top) handle.

Handwheel handle On modern taps, the cover and handle are replaced by a shrouded head that forms a handwheel. Shrouded heads not only give a neater appearance but also prevent water from wet hands going down the spindle and allowing detergent to wash the grease out of the tap mechanism.

High-neck taps The spout on an ordinary tap is about 22mm above its base, whereas the spout on a high-neck tap will be at least 95mm above its base. With a shallow sink this allows a bucket to be filled or large pans to be rinsed with ease. High neck taps are available with capstan, handwheel and lever handles.

Lever handle Another type of shrouded head has a lever handle, which is easy for elderly or disabled people to use as it can be pushed rather than gripped. Most lever-handle taps have ceramic discs so require only a quarter turn of the handle.

Hose union bib cock

A tap with a horizontal inlet now used mainly outdoors or in the garage. Most bib taps have an angled head and threaded nozzle, suitable for use with a garden hose.

MIXER TAPS

Two taps with a common spout are known as a mixer. The taps are linked either by a deck block (flat against the surface) or a pillar block (raised). Most mixers are two-hole types that fit into a standard two-hole-sink: one for the hot tap and one for the cold. Some mixers, however, need three holes (a centre hole for the spout) and some (monobloc types) only one.

Monobloc mixer
A single-hole monobloc mixer tap has a compact body with the handles and spout close together. Some designs have very narrow inlet pipes. There are monobloc designs for kitchens and bathrooms. Some kitchen monobloc mixers include a hot-rinse spray and brush fed from the hot water pipe by a flexible hose, so that the spray can be lifted from its socket for use.

Kitchen mixer The spout has separate channels for hot and cold water. This is because the kitchen cold tap is fed direct from the mains, and it is illegal to mix cold water from the mains and hot water from a storage cistern in one filling. The spout usually swivels and should be able to reach both of the bowls in a double sink. Kitchen mixers are available with capstan, handwheel and lever handles.

Bath/shower mixer Bath/shower mixer taps have a control knob that diverts the water flow from the spout to the shower handset. It will not provide a forceful spray but is a convenient addition to a bath.

Bath or basin mixer Hot and cold water merge within the mixer body, as both taps are usually fed from a cistern. It is illegal to fit this type of mixer on a fitting where cold water is supplied from the rising main and hot water comes from a cylinder. This is because, if mains pressure alters, differences in pressure might result in stored water being sucked back into the mains, and create the possible risk of contaminating drinking water supplies.

Choosing a washbasin and bidet

There is a huge choice of washbasins available, from a traditional ceramic pedestal to an ultra-modern polished limestone bowl. Pedestals hide pipework and give some support to the weight of the bowl. Wall mounted sanitaryware can look sleek, but needs a solid wall and strong fixings.

There is no standard size for basins – most are around 550mm wide and 400mm deep (from front to back). Order taps when you order the basin or bidet to ensure that your chosen unit has the right type of holes for the fittings.

Countertop basin These are often made in acrylic or enamelled pressed steel. They usually rest on the edge of a cutout or recessed ledge in the worktop and may have a rubber sealing ring and securing clips or be secured with mastic or sealant. Undermounted basins are screwfixed to a solid worktop from beneath.

TYPICAL WALL HUNG BASIN FIXINGS

The brackets for wall hung basins bear a heavy load and should only be fixed to a solid wall. If you have a stud wall, spread the weight with a countertop, washstand or pedestal type instead. Specialist fixings are usually supplied with the basin.

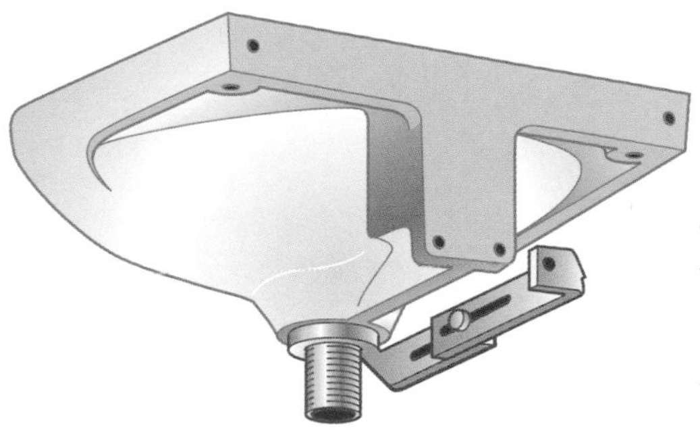

Wall plate and waste bracket The waste outlet of the wall-hung basin fits through the bracket, with its back nut below the bracket.

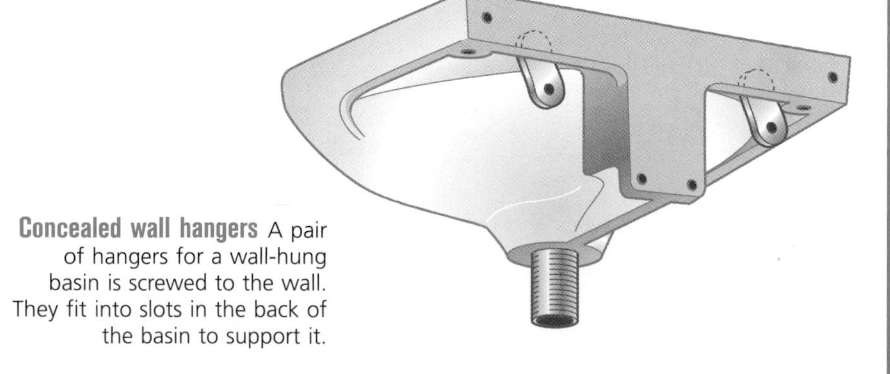

Concealed wall hangers A pair of hangers for a wall-hung basin is screwed to the wall. They fit into slots in the back of the basin to support it.

Pedestal basin Usually made of vitreous china but glass versions with stainless steel pedestals are also available. Luxury models can be much bigger. The pedestal hides the plumbing and helps to support the basin, but it is not the basin's sole support: the back of the basin must be screwed or bracketed to the wall. The basin may be joined to its pedestal with fixing clips or mastic. The pedestal is screwed to floor through holes at the back. The basin's height – usually 800mm – is not adjustable.

Semi-pedestal basin Usually made of vitreous china but glass versions with stainless steel pedestals are also available. The semi pedestal hides the plumbing but leaves the floor free and makes cleaning easier. The bowl is fixed in the same way as a pedestal basin.

Washstand basin Wall mounted washstand basins are usually of generous proportions and come with a surround supported on two legs so they appear freestanding. They are usually made of vitreous china or glass, with stands made of wood or chrome. The plumbing is not concealed.

Wall hung basin (below) Usually made of vitreous china, glass and stainless steel, wall hung basins are available as small as 400 x 350mm or as compact corner basins and space-saving short projection designs. They can be fixed at any height (low in a nursery, for example). Plumbing is visible unless boxed in.

Stand-alone basin These are designed to look like old-fashioned washbowls. They can be made of vitreous china, glass, stainless steel, wood and limestone. Plumbing for the basin is concealed in the unit on which it stands. Taps are fixed to the bowl or to the wall behind the basin.

Over-rim bidet (left) Usually vitreous china and typically 560–600mm deep and 350–395mm wide. The rim height of floorstanding models is about 400mm. Wall hung designs are also available and are especially suitable for wet rooms. Screwed to the floor inside the rear of the pedestal. Wall-hung types (for concealed plumbing) are bolted at the rear through wall to metal support brackets.

Rim-supply bidet Water enters under rim, so warm water warms the rim; a control diverts water to an ascending spray. This is more expensive than an over-rim bidet, and must be installed in accordance with Water Authority requirements. It is not suitable for DIY installation.

Replacing a bidet or washbasin

The operations involved in replacing a washbasin or bidet are very similar. You may need to make slight adaptations to the existing plumbing to accommodate the new appliance.

Before you start You need to turn off the water supplies to the taps (page 29), and disconnect (or cut through) the supply and waste pipes. Then you can remove the fixings holding the basin (and its pedestal, if one is fitted) or bidet in place. Fit taps, wastes and overflows to the new basin or bidet before installing it.

Tools *Basin wrench; spanner; long-nose pliers; steel tape measure; spirit level; damp cloth; screwdriver; bucket. Possibly also hacksaw.*

Materials *Basin or bidet; taps or mixer with washers; deep seal or anti-siphon bottle trap (page 61); waste outlet with two flat plastic washers, plug and chain; silicone sealant; fixing screws for wall (and floor if required); rubber washers to fit between screws and appliance. Possibly also flexible pipe with tap connectors.*

1 Fit the new taps or mixer. Place the sealing washer on the tap tail first, position the tap and screw on the backnut to secure the tap in place. Tighten it with a spanner. Check that single taps are correctly aligned.

HELPFUL TIP

If you want to renew the taps on an existing basin, it is often easier to cut through the supply and waste pipes and remove the basin from its supports. Even with a basin wrench, the back nuts can be extremely difficult to undo. With the basin upside down on the floor, it is easier to apply penetrating oil, and also to exert enough force without damaging the basin.

If removing the basin is not practicable, the tap handles and headgear can be replaced with a tap conversion kit, sold in packs with fitting instructions.

2 If you cut the supply pipes, fit flexible pipes (corrugated copper, braided hose or plastic) to each tap. These have tap connectors on one end and may include a service valve.

3 Attach the new waste outlet. Fit one sealing washer between the outlet and the appliance, then insert the outlet in its hole. Fit the second sealing washer from below and tighten the backnut. Use pliers to hold the outlet grid and stop the waste outlet from rotating as you do this. Ensure that the slots in the outlet tail line up with the outlet of the built-in overflow duct.

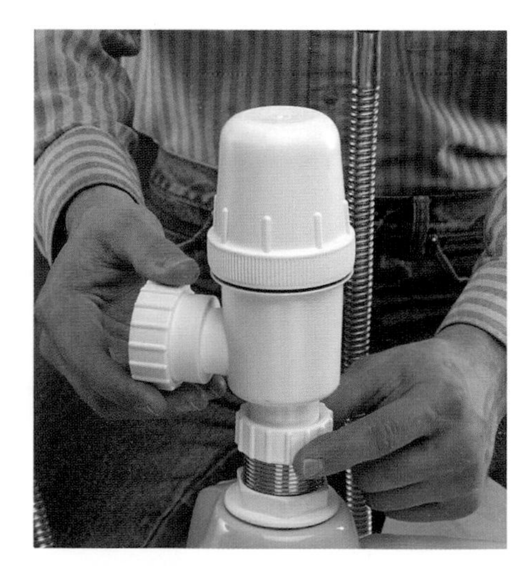

4 Fit the bottle trap to the tail of the waste outlet.

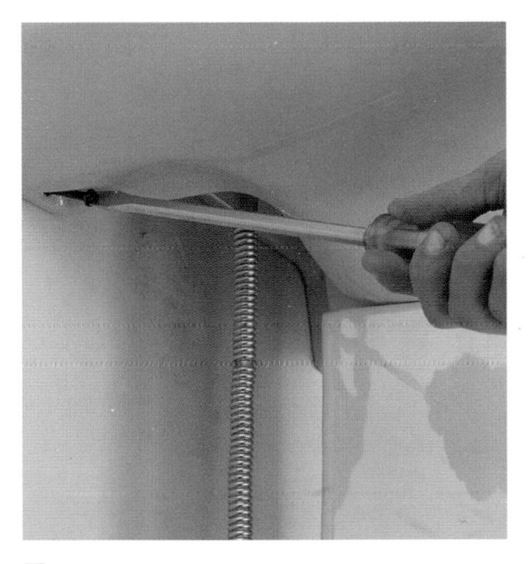

5 Set the new appliance in position and mark where new wall and floor fixings will be needed. Drill and plug the wall (and drill pilot holes in the floor too if necessary) and fix it in position.

6 Connect the taps or mixer to the supply pipes. If you disconnected the old tap connectors, there may be enough play in the pipes for you to reattach them directly to the new tap tails. If there is just a small gap, fit a tap tail extender to each tap and attach the old tap connectors to the extenders. If you cut the supply pipes, link the flexible connectors you attached to the taps in step 2 to the supply pipes using compression fittings. Include a service valve on each pipe (see right) if none is fitted.

7 Connect the outlet at the base of the bottle trap to the waste pipe.

8 Restore the water supply and check all joints for leaks. If necessary, tighten them.

9 Run a bead of silicone sealant around the appliance where it touches the wall.

Connecting a washbasin If it is difficult to disconnect the water supplies to the taps on the old basin, cut through the pipes lower down and use flexible connectors to connect the new taps to the supply pipes. Fit a service valve between the connectors and the pipes at the same time so that you can isolate the taps easily for future maintenance.

An over-rim supply bidet is connected to the supply pipes in the same way as a basin. You will need reducing fittings to connect the narrow tails of a monobloc mixer to 15mm diameter water supply pipes.

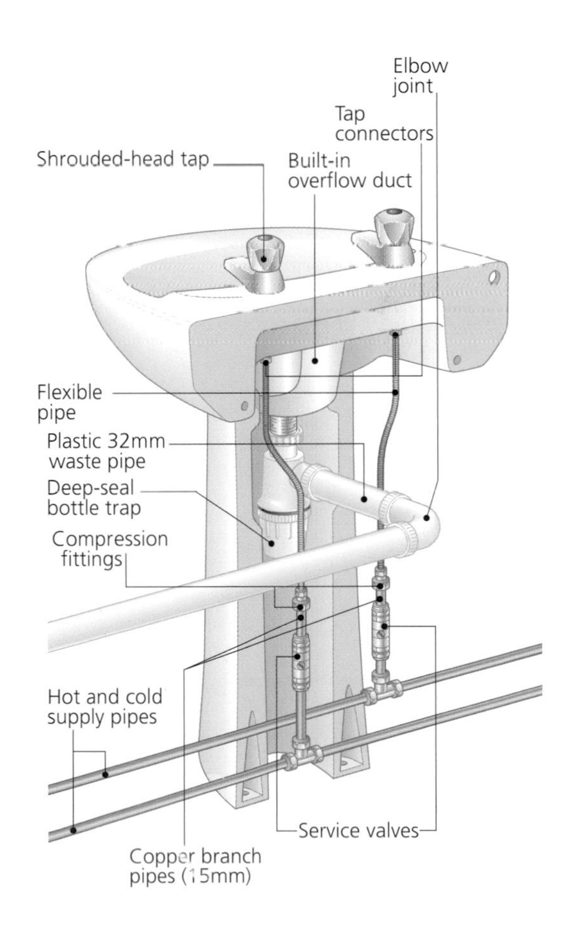

Installing a new countertop basin

Washbasins, particularly countertop basins on a vanity unit, can be installed in a bedroom to ease the demand on a family bathroom.

Before you start Give the local authority Building Control Officer details of your proposed arrangements for the new waste pipe connections.

1 Choose a site for the basin as near as possible to the bathroom waste pipe and supply pipes – ideally against a wall adjoining the bathroom.

2 Work out how to route the waste water from the basin to an existing waste pipe or direct to an outside drain (see opposite). Trace the routes of existing hot and cold supply pipes and work out the shortest route possible for the new pipe.

3 Check that the basin position will allow sufficient space for a person to use it. Generally, allow at least 640mm bending room in front of the basin, and at least 300mm elbow room at either side.

4 Check that the installation of the basin will not interfere with any electric cables, gas pipes or other fittings, especially where you need to make a hole in a wall.

5 Fit the taps, waste outlet and trap to the new washbasin as described in Replacing a bidet or washbasin (page 68). Use a deep-seal trap (or an anti-siphon trap, if required).

6 Cut off the water supply (page 29) and use tee connectors to run hot and cold-water pipes to the basin site as 15mm branch pipes from the supply pipes to the bathroom. Do not tee into the pipes supplying a shower, unless it has a thermostatic mixer (page 85). If you find you have to tee into a 22mm distribution pipe, you will need an unequal tee with two 22mm ends and a 15mm branch.

7 Fit the waste pipe in position. If you have to make a hole through the wall, do it as for Installing an outside tap (page 91). If you plan to connect the new waste pipe into an existing one, insert a swept tee joint and link the new pipe to it.

8 Fit the basin or basin unit to the wall and connect it to the supply pipes and waste pipe (see previous page).

Routeing the waste pipe

The waste pipe run should be no more than 1.7m long. It must slope enough for the water to run away – not less than 20mm for each 1m of run – but the depth of fall to the pipe outlet should be no greater than about 50mm for a pipe under 1m long, or about 25mm for a longer pipe. If you cannot avoid a pipe run longer than 1.7m, prevent self-siphonage by fitting an anti-siphon bottle trap (page 61).

Alternatively Use a waste pipe of larger diameter – 40mm instead of 32mm. This pipe run should be no longer than 3m, and you will need a reducer fitting to connect

HELPFUL TIPS

If you tee into an old water pipe, it is likely to be an imperial size of $^{3}/_{4}$in internal diameter. There is no way of recognising an imperial size, except by measuring the internal diameter of the pipe once you have cut into it.

With $^{3}/_{4}$in copper pipe, you can fit 22mm compression joints straight onto it provided that you substitute the olives for $^{3}/_{4}$in olives, available at a plumbers' merchant. You can get metric/imperial adaptors for most types of plastic push-fit fittings.

On soldered capillary joints, use adaptors, which are available from a plumbers' merchant, that convert $^{3}/_{4}$in piping to 22mm, before inserting the tee.

¾ in to 22mm adaptor
¾ in supply pipe
22mm tee joint

Flexible plastic push-fit joints are useful for fitting between the trap and waste pipe where alignment is difficult – for example, round a timber stud in a partition wall.

40mm pipe to the 32mm trap outlet. If you need a run longer than 3m, ask the advice of your local authority Building Control Officer.

Linking to an outside drain

How the waste pipe is linked to a drain depends on the household drainage system. If you are unable to link into the waste pipe from the bathroom basin, fit a separate waste pipe through an outside wall to link with a household drain.

If the bedroom is on the ground floor, direct the waste pipe to an outside gully, if possible, such as the kitchen drain. The pipe must go below the grid.

Where the bedroom is on an upper floor, the method of connecting the waste pipe depends on whether you have a two pipe or single-stack system (page 17).

With a two-pipe system, you may be able to direct the waste pipe into an existing hopper head. Alternatively, you can connect the waste pipe to the waste downpipe (not the soil pipe) using a stack connector.

In a house with a single-stack system, the waste pipe will have to be connected to the soil stack. For this you need the approval of the local authority Building Control Officer. There are regulations concerning the position of the connection in relation to WC inlets, and the length and slope of the pipe is particularly critical. The connection is made by fitting a new boss, which is a job best left to a plumber.

Bedroom basin installation

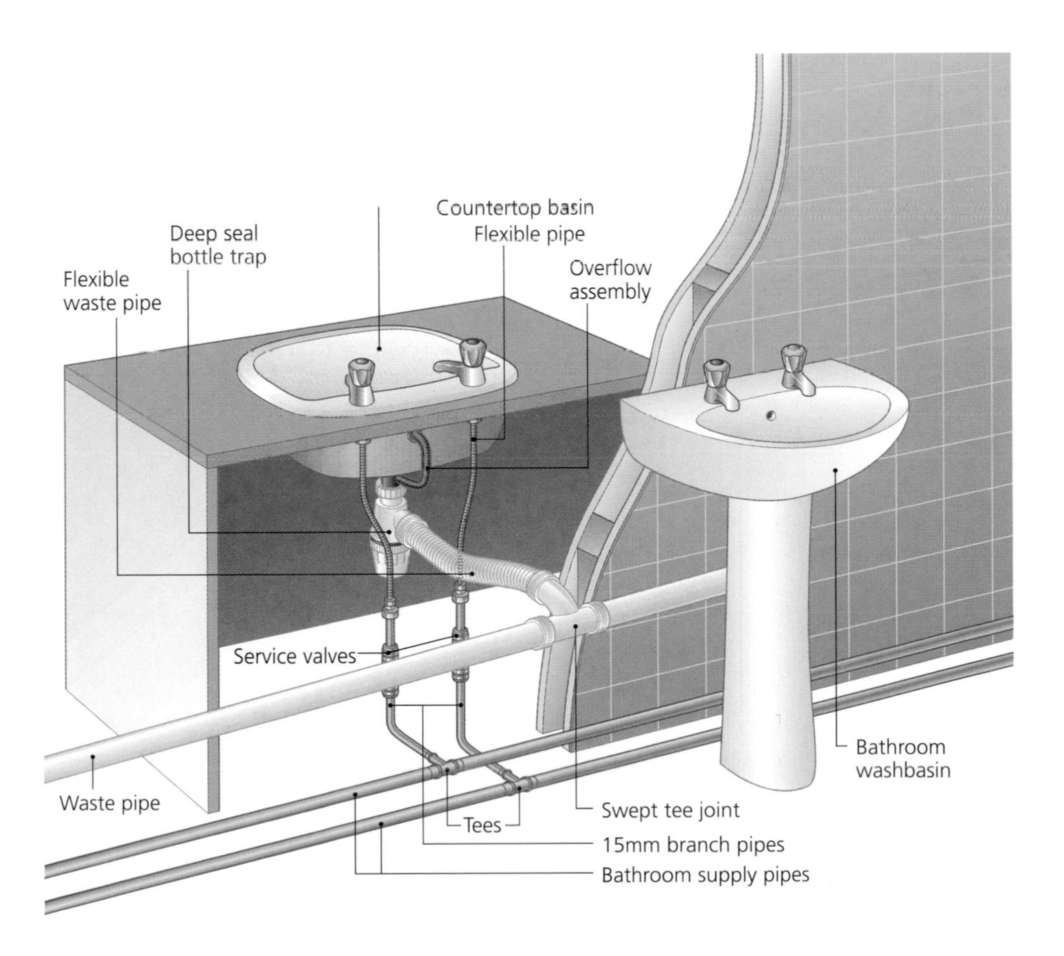

Renovating and repairing a bath

Major bath renovation is best done by professional firms. They take a day or two and will repair cracks or chips and apply a new coating, available in a range of colours. You can do smaller repair jobs yourself, using a DIY kit.

Repairing bath enamel

Before you start Remove every trace of grease or flaky enamel from the surface of the bath. Once clean, do not touch the surface again with your fingers.

1 Fill small surface chips with an epoxy resin filler (such as car-body filler) or use one of the proprietary bath repair kits available on the market.

2 Rub the surface smooth with fine abrasive paper, then coat the repaired area with two coats of matching bath enamel, following the instructions on the pack. Touch-up sticks are available in a variety of colours to cover small repairs.

Repairing a plastic bath

Burns on an acrylic surface cannot be repaired. However, chips can be filled with a special acrylic repair paste available as a kit. You can polish out scratches with metal-polish wadding because the colour goes right through the material.

However, glass-reinforced plastic (GRP) can be damaged by abrasive cleaners as only the top layer of material is coloured.

Changing bath taps

Because of the cramped space at the end of a bath, fitting new taps to an existing bath can be difficult. It is often easier to disconnect and pull out the bath (see right) so that you have room to apply enough force to undo the backnuts of the old taps.

The alternative to disrupting the bathroom is to fit new tap headgear only, using a tap conversion kit.

Removing an old cast-iron bath

A cast-iron bath may weigh around 100kg, so you will need helpers to move it. A pressed-steel bath is lighter. It can usually be moved intact.

Before you start Unless you want to keep a cast-iron bath intact, it is easier to break it up after disconnecting it than to remove it whole. Be careful when you break it up, as the pieces are often jagged and very sharp.

Tools *Torch; adjustable spanner; safety goggles; ear defenders; club hammer; blanket; protective gloves. Possibly also padded pipe wrench; screwdriver; hacksaw.*

1 Cut off the water supply (page 29).

2 Remove any bath panelling. It is often secured with dome-head screws, which have caps that cover the screw slot.

3 With a torch, look into the space at the end of the bath to locate the supply pipes connected to the tap tails, and the overflow pipe. In older baths, the overflow pipe is rigid and leads straight out through the wall. In more modern types the overflow is flexible and connected to the waste trap.

4 Check the position of the hot supply pipe: it is normally on the left as you face the taps. Use an adjustable spanner to unscrew the tap connectors from the supply pipes and pull the pipes to one side. If unscrewing is difficult, saw through the pipes near the ends of the tap tails.

5 Saw through a rigid overflow pipe flush with the wall.

6 Disconnect the waste trap from the waste outlet. For an old-style U-bend, use an adjustable spanner. A plastic trap can normally be unscrewed by hand, but use a padded pipe wrench if it proves difficult. Pull the trap to one side. Disconnect a flexible overflow pipe from the overflow outlet.

7 If the bath has adjustable legs – normally brackets with adjustable screws and locking nuts – lower it to lessen the risk of damaging wall tiles when you pull it out. But if the adjusters on the far side are difficult to reach, lowering may not be worth the effort.

8 Pull the bath into the middle of the room ready for removal or break-up.

9 To break up the bath, drape a blanket over it to stop fragments flying out, and hit the sides with a club hammer to crack the material into pieces.

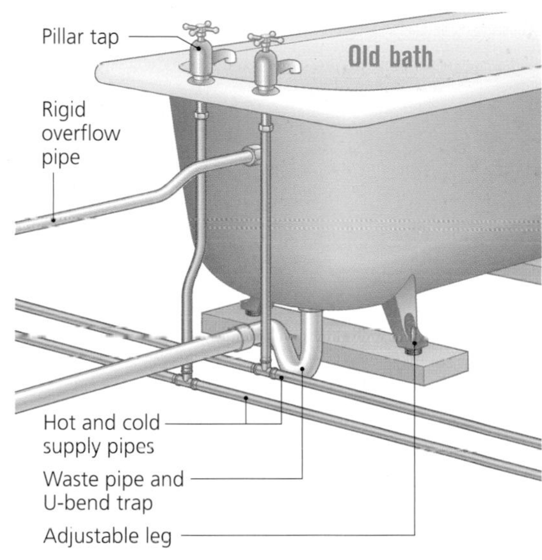

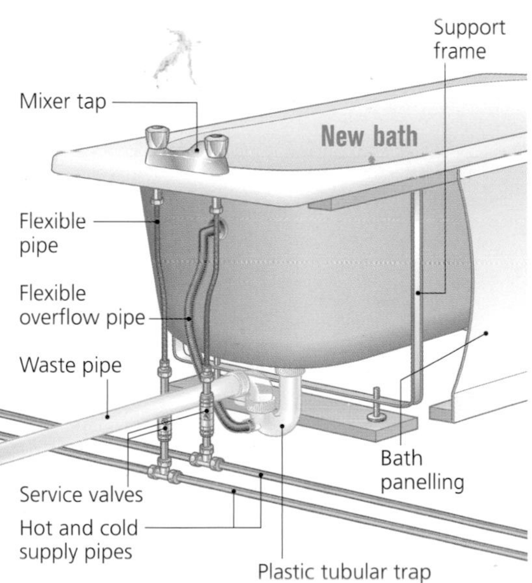

Connections to an old and new bath As when replacing a washbasin or bidet (page 68), add a service valve on each of the supply pipes before connecting them to the new taps. This allows them to be isolated easily for future maintenance.

Installing a new bath

This is a good time to re-think your bathroom: you can fit the new bath in a different position and take advantage of the latest designs.

Before you start Assemble as many fittings as possible onto the new bath before you remove the old one. Not only will fitting be easier before the bath is in position, but the water will not be cut off for as long.

Use flexible connectors and a flexible waste joint. If you want to put the bath in a different position, you will have to work out how to re-route the waste pipe, as well as adapting the supply pipes.

Tools *Two adjustable spanners; spirit level; damp cloth. Possibly also long nosed pliers; small spanner; hacksaw; screwdriver.*

Materials *Bath; two 25mm thick boards to support its feet; two new taps or a mixer tap (with washers); two 22mm flexible pipe connectors; 40mm waste outlet with plug and two flat plastic washers; bath trap (deep-seal if the waste pipe links to a single stack) with flexible overflow assembly; silicone sealant; PTFE tape.*

---

1 Fit the supporting frame following the maker's instructions. It is usually done with the bath placed upside down.

2 Fit the taps or mixer and the tap joints in the same way as for a washbasin. Some deck mixers come with a plastic gasket to fit between the deck and the bath. On a plastic bath, fit a reinforcing plate under the taps to prevent strain on the bath deck.

3 Fit the waste outlet. This may be a tailed grid fitted in the same way as for a sink. Or it may be a flanged grid only, fitted over the outlet hole (with washers on each side of the bath surface) and fixed with a screw to a tail formed at one end of the flexible overflow pipe.

4 Fit the top end of the overflow pipe into the back of the overflow hole and screw the overflow outlet, backed by a washer, in position.

5 Slot the banjo overflow, if supplied, onto the threaded waste outlet and fit the bottom washer and back nut. Attach the trap, then fit the bath into position with a flat board under each pair of feet in order to spread the load.

Choosing a new bath

The style and shape of the bath you choose is based on personal preference. But it helps to have a clear idea of what you are looking for when you go to choose one.

Type Many people are happy with a standard, traditional bath, but there are exciting new types of bath available that are worth considering.

• Whirlpool systems have water jets in the side of the bath, which give the bather an invigorating massage.

• Spa baths have air nozzles in the base of the bath, forcing bubbles upwards and creating a stimulating wave action.

• Hydro baths combine whirlpool side jets with spa nozzles in the base. In addition they may offer jets to stimulate the neck and shoulders and the feet.

Material Most modern baths are made from acrylic or glass-reinforced plastic (GRP), and are light and quite easy to install. They are usually cradled in a support frame to avoid distortion when filled.

• Vitreous-enamelled pressed-steel baths are lighter and cheaper than old-style, porcelain-enamelled cast-iron baths.

Quality Cheaper baths are thinner and tend to need more support to prevent sagging and creaking when occupied. A good-quality plastic bath should be at least 6mm thick.

• Pressed-steel and cast-iron baths are strong and firm, but can feel cold, often initially lowering the temperature of the water until they warm up. Enamel is easily chipped, and often requires special cleaning materials.

Shape Bath shapes may also include two-person, corner-fitting and circular. They can be free-standing, with ornate legs, or fitted into a panelled framework. Sunken baths are glamorous but can be rather difficult to clean.

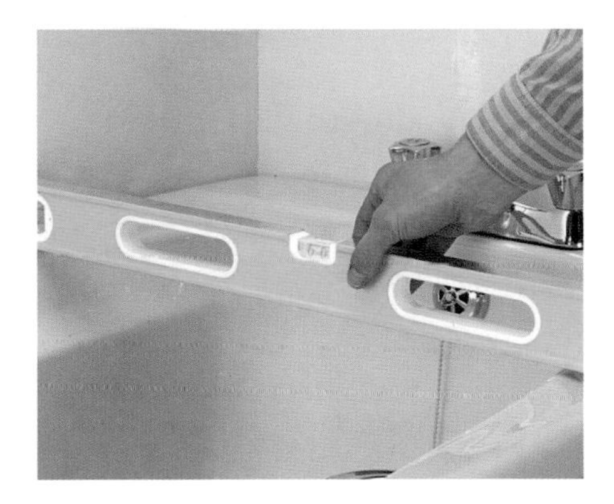

6 Place a spirit level on each of the four sides of the bath to check that it is horizontal. If necessary, adjust the legs until the bath is perfectly level. Then tighten the locking nuts on the adjustable legs.

7 Fit the flexible connector on the farthest tap to its supply pipe, making a compression joint. If the supply pipe is too high, cut it back to a convenient length, leaving it too long rather than too short, as the connector can be bent slightly to fit.

8 Connect the second tap tail in the same way as the first. Then connect the trap outlet to the waste pipe (normally a push-fit joint). Restore the water supply and check the joints for leaks. Tighten if necessary, but not too much.

9 Fix the bath panels according to the maker's instructions. They may screw or clip to a wooden frame, or be fixed to a batten screwed to the floor. Fill the bath with water before sealing the joints between the bath sides and walls with silicone sealant. This ensures that the bath will not settle in use and pull the sealant away.

HELPFUL TIPS

If, on a mixer, the hot and cold indicators are on different sides from the appropriate supply pipes, reverse the discs on the tops of the taps or cross the flexible pipe connectors to join the taps to the correct pipes. The hot tap should be on the left.

Types of WC suite

A WC suite consists of a cistern and a pan. The cistern can be low or high level, joined by a flush pipe to the pan, or close-coupled with a direct connection. Modern suites are designed to use less water than old ones – 6 litres rather than 7.5 or 9 litres. When siting a WC suite, allow at least 530mm of space in front of the pan, and for about 760mm overall space from side to side.

Types of cistern

Cisterns are made of plastic or vitreous china (ceramic), and screwed to the wall from inside, through the back. The flushing control is a central push button or a side lever (fitted to right or left of the cistern).

For the flushing action to be correct, the cistern must be at the height given by the manufacturer – usually with the base about 600mm above the floor. Water flow into the cistern is controlled by a ballvalve. An older cistern fed from a cold water tank is quieter than one fed from the rising mains.

For more about the types of ballvalves (and push-button flushes), see page 46 and 47.

Standard cistern A direct-action cistern installed at a low level is the standard option. It can also be used at high level, and is quieter than the chain-pull type. If a suite is converted from high to low-level, the standard size cistern – about 200mm from front to back – is too deep to go behind the pan, so a slimline type is needed.

Slimline cistern Measures the same as a standard cistern from side to side but as little as 115mm from front to back. However, it still provides a full flush. It may be concealed behind panelling.

Dual-flush type A variation on the direct-action cistern is the dual-flush type, which provides either a water-saving small flush, or a double-sized flush for solid waste.

Low-level suite

In a low-level suite, the cistern is either linked to the pan by a short flush pipe or is close-coupled, with the cistern and pan joined together in one unit.

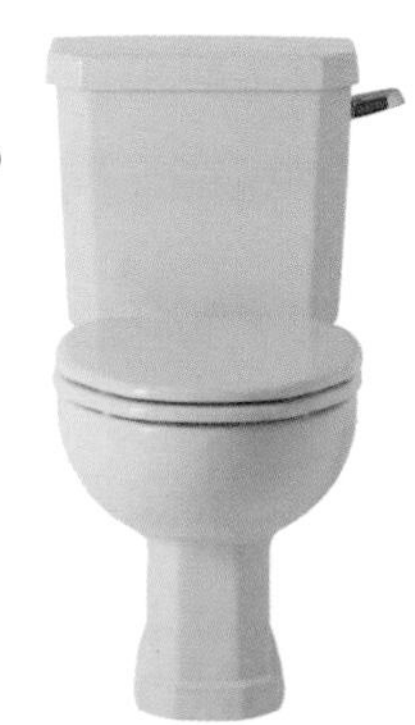

Standard cistern

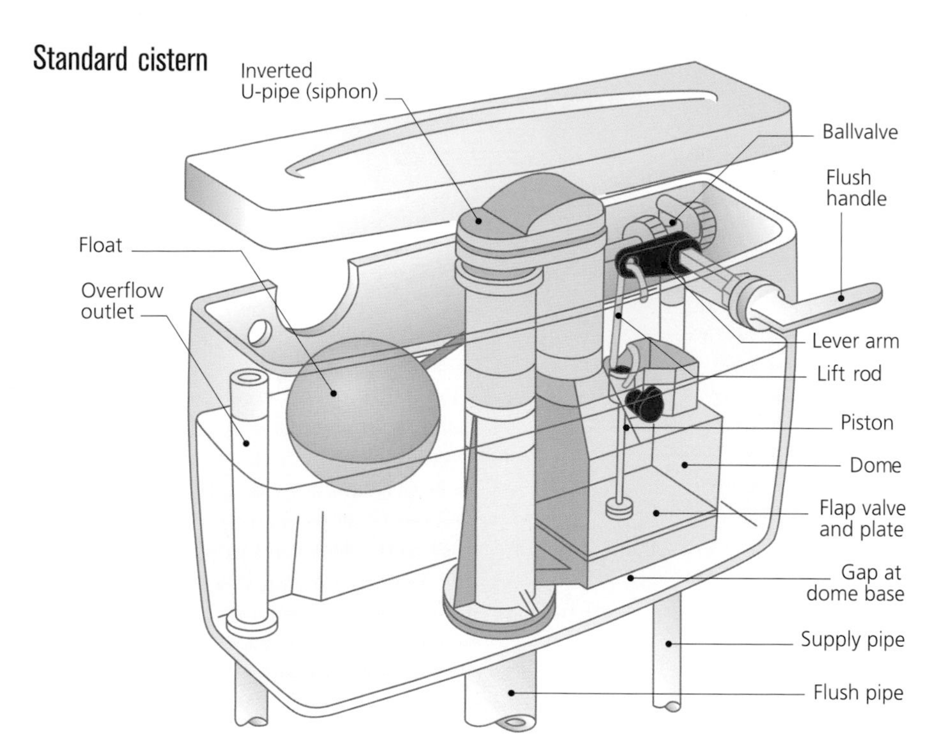

HOW A CISTERN WORKS

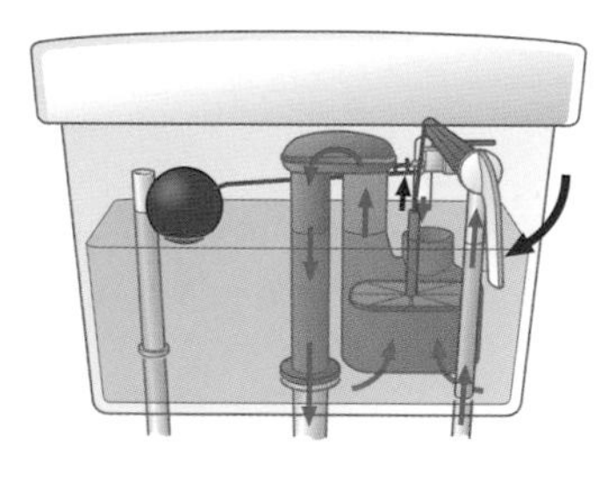

1 An inverted U-pipe in the cistern is linked to the flush pipe into the pan, and at the other end opens out into a dome (siphon). When the flush is operated, a lift rod raises a plate in the dome and throws water into the crown of the U-bend.

2 Openings in the plate are covered by a plastic flap valve held flat by the weight of the water. As water falls down the flush pipe, it creates a partial vacuum, causing water to be sucked up through the plate openings and raising the flap valve.

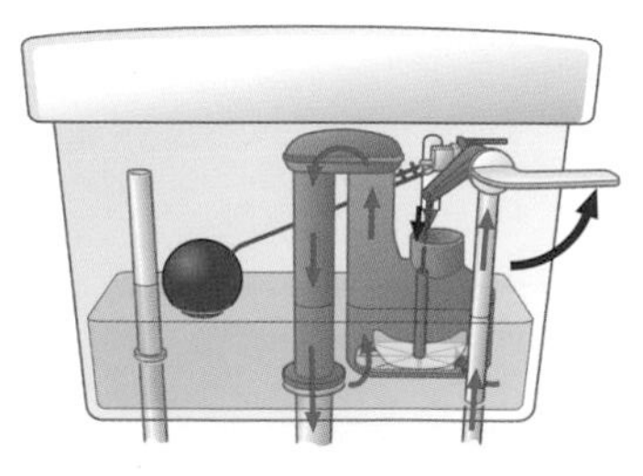

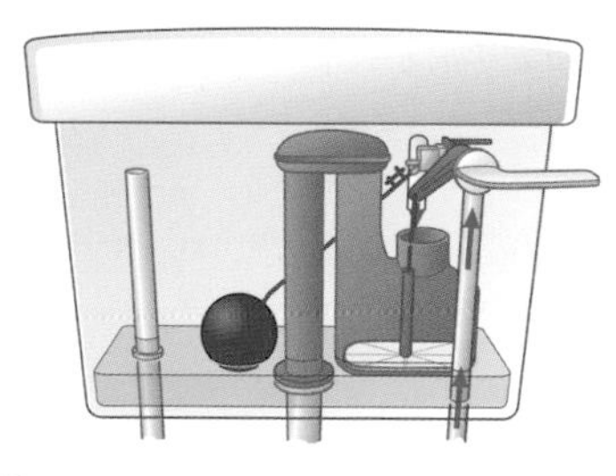

3 The base of the dome is about 10mm above the cistern base. When the cistern water level falls below the dome base, air drawn in breaks the siphonic action and stops the flush.

On a dual-flush cistern the two-part button allows the user to select a reduced amount of water for the flush. This system works by means of a hole in the side of the dome. If the flush control is released immediately, the hole lets in air to break the siphonic action after approximately 4 litres of water has been siphoned. When the control is held down for a few seconds, it operates a device that plugs the air hole, allowing a double flush.

Types of WC pan

Pans are normally made from vitreous china and are screwed to the floor through holes in the base of the pedestal. Some fit flush to the wall and others are wall-hung.

Unless connected directly to an internal soil stack, the outlet pipe of the pan needs to be connected to a soil pipe that passes out more or less horizontally through the wall (usual for upstairs WCs) or passes vertically through the floor (common for downstairs WCs).

There are two types of pan: wash-down pans and double-trap siphonic pans.

Wash-down pan

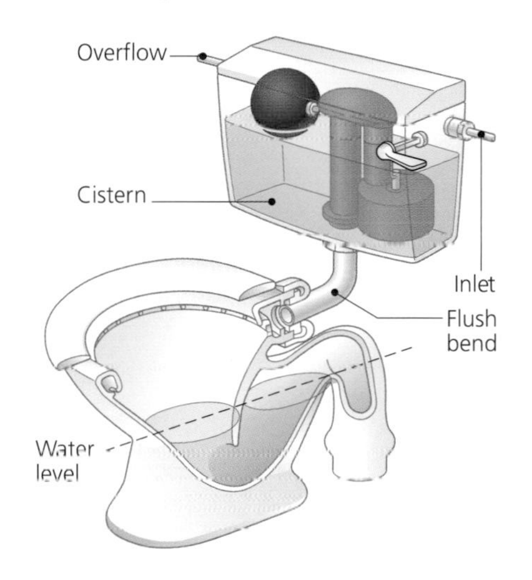

All new pans sold today are of the wash-down pan type, where the pan is cleared by water from the cistern splashing on the sides of the pan and forcing the contents of the pan through the trap, leaving clean water in the bottom of the pan.

Three designs of pan are made: horizontal outlet; P-trap (just below the horizontal) and S-trap (vertical). The vast majority of pans now being sold have a horizontal outlet, but this can be connected to a soil pipe formerly fitted to a vertical outlet using an angled joint (overleaf).

Double-trap siphonic pan

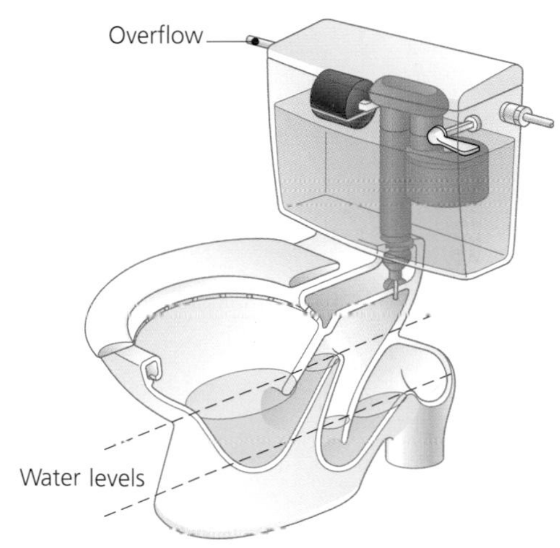

Some existing pans are the double-trap siphonic type; these are no longer made, though spares should still be available. When the cistern is flushed, the pan starts to empty by suction before the flush water reaches it, making it very quiet and efficient.

MACERATOR UNITS

A macerator unit allows you to put a WC in a room distant from the main soil stack. It takes waste from the WC pan and passes it through a shredder so that it can be carried away through a small-bore (22mm or 32mm) plastic pipe. The WC pan can be 20–50m from the soil stack if the pipe is horizontal, but pumping distance and height are interlinked, so pumping vertically would restrict the horizontal distance.

The installation of a macerator unit must meet the Building Regulations requirements, so you should inform a local Building Control Officer if you intend to install one. You will also need an electrical connection to drive the shredding motor and this must be via a fused connection unit (unswitched or via a flex outlet if within reach of anyone using the bath or shower).

There are a wide variety of macerator units available: the simplest is for use only with a WC; the most complex will take a whole bathroom suite (WC, wash basin, bidet and bath or shower). High-capacity units are available to take the output from a power shower. There are also macerator units designed for kitchen waste (sink, washing machine and dishwasher). Macerator units can be bulky, but there are slimline units available that fit neatly behind a conventional WC pan, below the cistern.

TYPES OF WC PAN JOINTS

Plastic push-fit joints are now universally used and come in a variety of shapes to allow connection of virtually any pan to any soil pipe. Most joints are either straight or 90° (for horizontal or vertical soil pipes), but offset joints, extension joints and even fully flexible joints are also available.

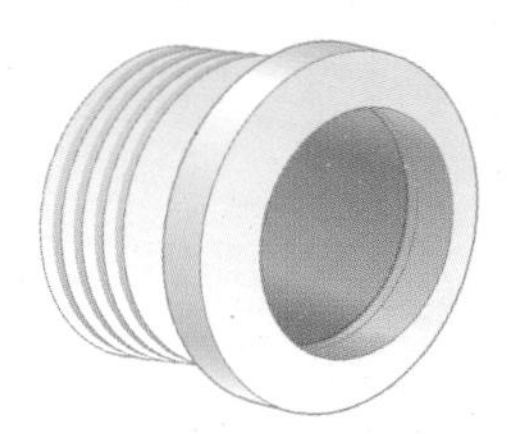

Straight push-fit pan joint For a straight link between the pan outlet and the inlet branch to the soil pipe. The cupped end fits over the pan outlet, and the narrow (spigot) end inside the soil-pipe inlet. Different diameters and lengths are made. Before buying, check the outside diameter of the pan outlet, the inside diameter of the soil-pipe inlet, and the distance to be bridged. Joints have watertight seals at each end. Offset types can be used where the alignment is not exact.

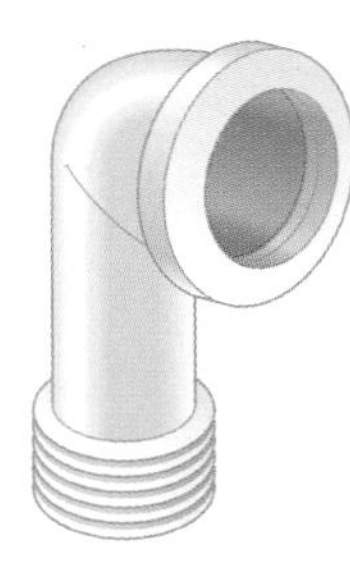

Angled push-fit pan joint A 90° joint for converting a horizontal (P-trap) pan outlet to a down-pointing (S-trap) outlet for a floor-exit pipe. It can also be used to link a horizontal outlet to a wall-exit pipe situated at right angles to the pan.

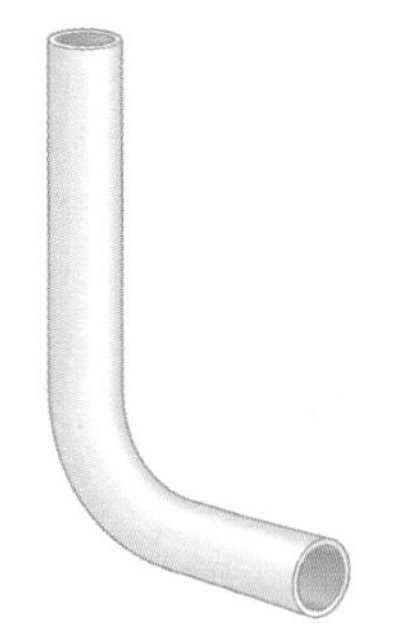

Flush pipe Angled plastic pipe linking a separate cistern to the WC pan. Pipes for high-level suites are normally 32mm in diameter, and pipes for low-level suites have 38mm diameters.

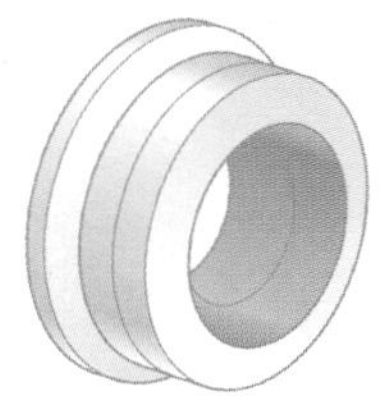

Rubber cone joint For linking the flush pipe from the WC cistern to the flush horn of the pan.

Replacing a WC pan

At one time, WC pans were always cemented to a solid floor, but the setting of the mortar often put a strain on the pan and caused the china to crack. Now they are usually screwed down to a wooden or a solid floor.

Before you start An old or cracked WC pan with a down-pointing outlet cemented to a floor-exit metal soil pipe is the most difficult type to remove. Examine yours carefully before attempting to remove it.

Tools *Screwdriver; spirit level. Possibly drill and wood or masonry bits; safety goggles; club hammer; cold chisel; rags; old chisel; thin pen, pencil or nail; trimming knife.*

Materials *WC pan and seat; pan fixing kit; rubber cone connector; suitable push-fit pan connector. Possibly also wall plugs (for a solid floor); packing (to steady the pan) such as wood slivers, vinyl tile strips or silicone sealant.*

Removing a pan with a horizontal outlet

1 Disconnect the flush pipe by peeling back the cone connector. Alternatively, chip away a rag-and-putty joint with an old chisel. Protect your eyes.

2 Undo any screws used to secure the pan to the floor.

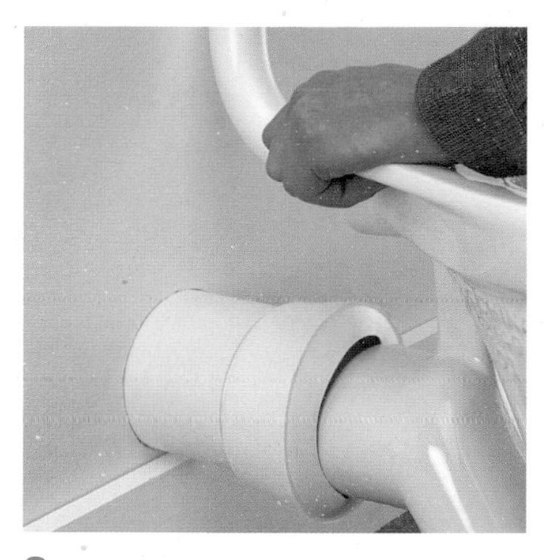

3 Pull the pan forward slowly, moving it from side to side, to free it from the soil pipe inlet. It should come away easily. If you have any difficulty, break the pan outlet in the same way as for a down-pointing outlet (below).

4 If the outlet joint was cemented with putty or mastic filler, chip it off the soil pipe inlet.

Removing a pan with a down-pointing outlet

1 Disconnect the flush pipe in the same way as for a horizontal outlet pan.

2 Undo the floor screws, or break cement with a hammer and cold chisel.

3 To free the pan outlet, put on safety goggles and use a club hammer to break the outlet pipe just above its joint with the drain socket in the floor. Then pull the pan forward, away from the jagged remains protruding from the soil pipe socket.

4 Stuff rags into the socket to stop debris falling in, then chip away the rest of the pan outlet with a hammer and cold chisel. Work with the chisel blade pointing inwards, and break the china right down to the socket at one point. The rest of the china should then come out easily.

5 The new push-fit connector fits directly into the pipe so the collar is redundant. Break out the collar gently by tapping it outwards and remove the pieces so the pipe finishes flush with the floor level. Alternatively cut it off with an angle grinder.

6 Clear away any mortar left where the pan was cemented to the floor, leaving a flat base for the new WC pan.

Fitting a separate pan

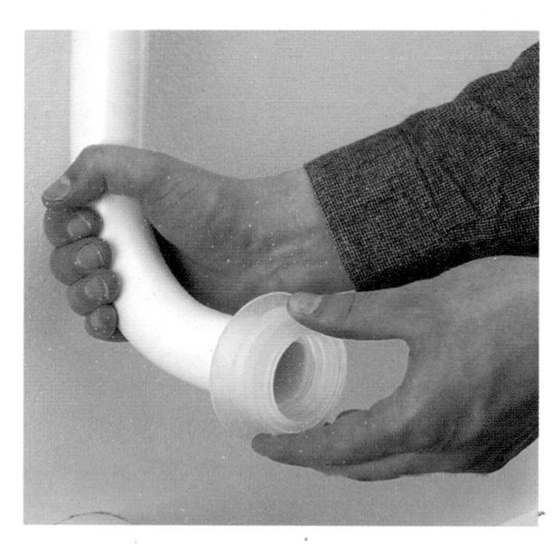

1 Fit a rubber or plastic cone connector to the flush-pipe outlet, unless one has already been fitted.

2 Fit a plastic push-on connector to the pan outlet. If the soil pipe socket for the pan is in the floor, use an angled 90° connector to link a horizontal pan outlet to the vertical inlet to the soil pipe.

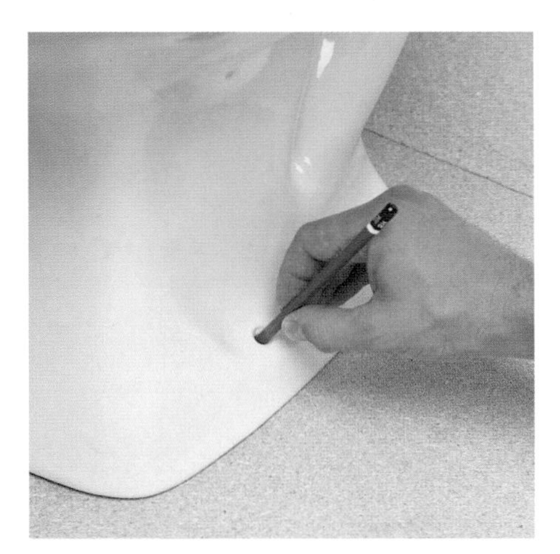

3 Mark the position of the holes in the pan on the floor, using a slim marker such as a ballpoint pen or a pencil.

4 With the pan still in position, draw a line round its base so that it can be put back in place accurately. Then remove it.

5 You need rust-proof screws to secure a WC pan to the floor. You can buy 'kits' which have the screws, wall plugs, plastic bushes and caps to cover the screw heads. On wooden floors, drill a pilot hole for the screws; on a solid concrete floor, use a masonry drill bit to make holes in the concrete and insert wall plugs.

6 If the soil pipe inlet is a floor socket, remove the rags, taking care not to spill any of the debris in the inlet.

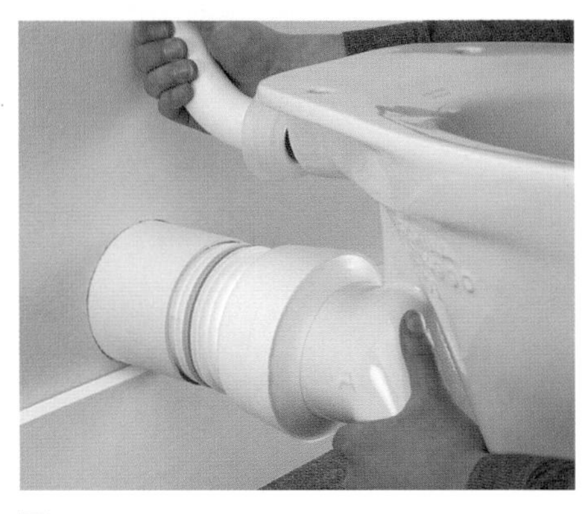

7 Carefully lift the pan into position using the previously marked outline to guide you, and at the same time positioning it so that you can push the flexible connector into the soil pipe inlet. Fold back the cone connector and slip it over the flushing horn of the pan.

8 Put in the screws to hold the pan firmly in position. However, do not tighten them fully yet.

9 Use a spirit level placed across the top of the pan to check that it is level from side to side and from front to back.

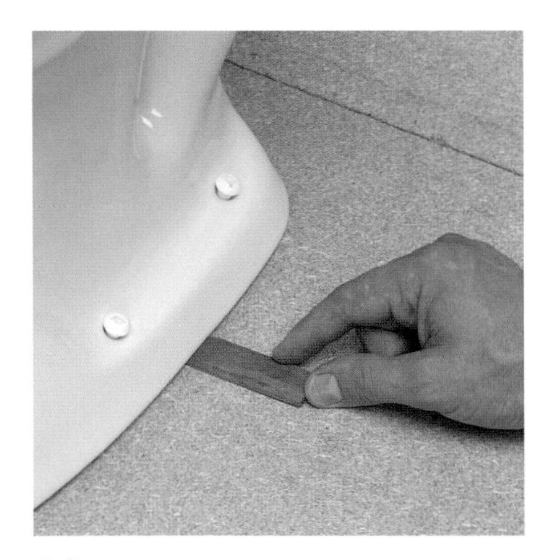

10 To level the pan, loosen the plastic nuts and pack under the pedestal with strips of vinyl tile, or use a bead of silicone sealant to steady the pan and provide an even bed. Once you are sure that the pan is level, screw it down firmly.

Fittings for a separate low-level suite

Most modern WC pans have a horizontal outlet. If the replacement pan has to be fitted to a soil pipe on a ground floor, connect the P-trap outlet of the pan to the soil-pipe inlet using an angled push-fit joint (page 78). After fitting a WC suite, you may find that condensation – particularly apparent on a ceramic cistern – is a nuisance and leads to damp walls and floors. Make sure the WC or bathroom is properly ventilated. In a bathroom, avoid drip-drying washing over the bath, as this contributes to condensation.

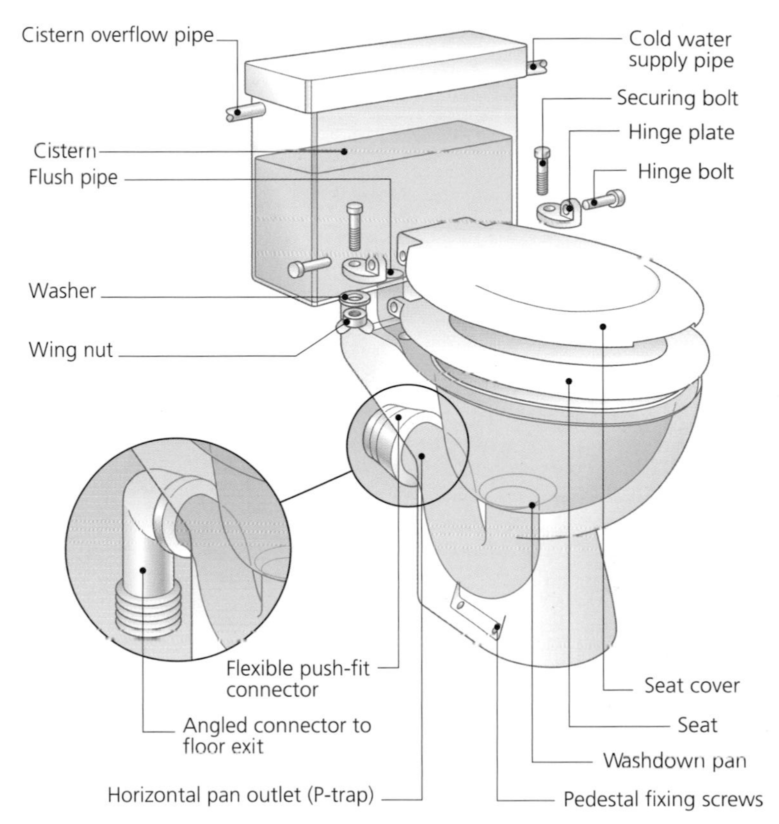

Checking the fitting

When fitting a new WC pan, make sure that it is level and that the connections to the flush pipe and soil pipe are true. Otherwise slow pan clearance or a blockage could result.

On an existing WC pan with old-style joints rather than flexible connectors, blockages can occur because the openings are out of true or partially obstructed by the jointing material. Putty from an old rag-and-putty joint between the flush pipe and horn may have squeezed into the flush inlet and be impeding the inflow of water.

Putty or other jointing material could be obstructing the joint between the pan outlet and the soil-pipe inlet. This is evident if the water rises slowly in the pan before it flows away.

Make sure the pan is firmly fixed and that the fixing screws have not worked loose. Check that the pan is horizontal when fixing it in place.

FITTING A NEW WC SEAT

A WC pan seat and cover are usually fixed onto hinge bolts or a rod at the back. These fit into hinge plates or covers at each side of the pan.

Hinge plates or covers are each held in place by a securing bolt that fits through a fixing hole in the back of the pan and is secured by a wing nut.

Make sure that you insert washers to shield the pan from the head of the securing bolt and the wing nut. Washers shaped to the pattern of hinge covers are often supplied. Screw the wing nuts firmly finger-tight.

A WC seat breaks easily if misused – such as if you stand on it to close a window. If you simply have to stand on something, lift the seat and balance yourself on the rim of the pan.

Planning a shower

A shower is a useful addition to any home, but you need to think about where and how to install it.

There are four main types of shower – bath/shower mixers, shower mixer valves, instantaneous electric showers and 'power' showers. All can be installed over an existing bath and all (except bath/shower mixers) in a separate cubicle.

Electric showers are connected directly to the rising main, so can be fitted in any house; power showers can only be fitted to stored hot and cold water supplies. For mixer showers, the ideal is to have both hot and cold water supplies at the same pressure – either stored or from the mains.

Water supply

The plumbing needed depends on the type of shower you fit and your water supply.

Bath/shower mixer No extra plumbing is normally required for a bath/shower mixer: it simply replaces the bath taps.

Shower mixer If you are installing a shower mixer valve with stored hot and cold water, a separate pipe should be taken from the cold water cistern for the cold supply and the hot supply should also be a separate pipe run as the first connection to the vent pipe above the hot water cylinder. These connections avoid the shower running hot (or cold) when water is drawn off elsewhere in the house (can also be avoided with a thermostatic mixer shower).

Where you have different pressures (cold from storage cistern, hot from combi boiler, for example), the cold needs its own supply from the cold water cistern and you must use a pressure-balanced shower mixer valve.

Note that you should inform your water supply company if you are making connections to mains-fed water pipes.

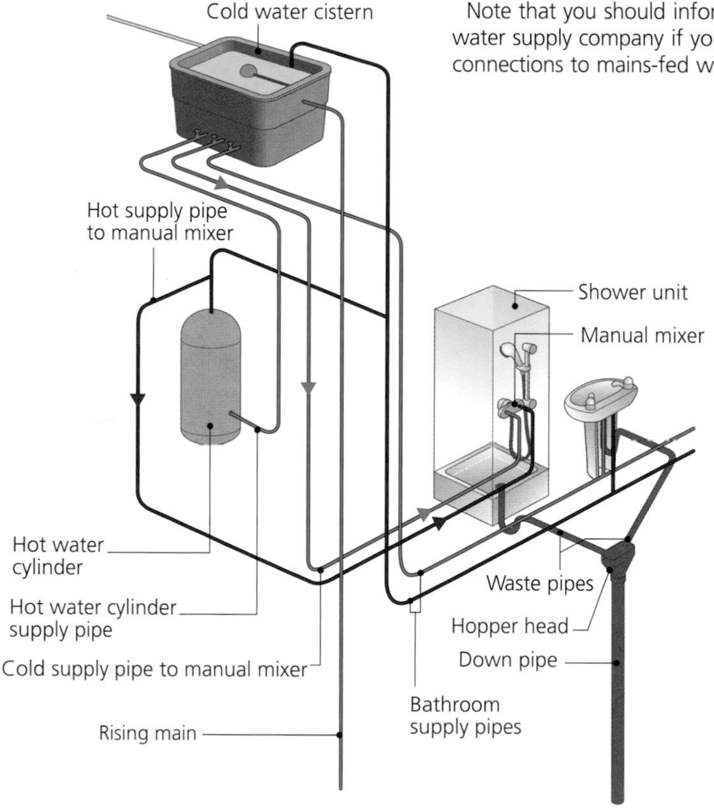

Pipework for a shower mixer For a shower mixer, take the cold supply pipe direct from the cold water cistern to avoid risk of scalding when other cold taps are turned on. Take the hot supply from the hot water cylinder distribution pipe – tee in above cylinder height. For a thermostatic mixer, which has a temperature stabiliser, you can tee in to bathroom supply pipes.

Electric shower This needs only a connection to the cold water rising main.

Power showers Because of the high volumes of water used, the cold supply should be direct from the cold water cistern and the hot supply may need a special connection to a hot water cylinder.

Drainage

Showers fitted over a bath need no extra drainage, though you might need a different trap with high-volume showers.

A separate cubicle will need its own drainage arrangements and you may need approval for this from your local authority Building Control Officer.

Achieving adequate pressure

For a mixer supplied from the household's stored hot and cold water supplies, the bottom of the cold water cistern needs to be at least 900mm (preferably 1.5m) above the showerhead for the flow to be adequate. If you do not have sufficient water pressure to supply a shower at the required position, there are two ways to increase it: you can either raise the height of the cistern or have a booster pump installed. Or you could fit a power shower.

Raising the cistern The cold water cistern can be raised by fitting a strong wooden platform beneath it, constructed from timber struts and blockboard or plywood. You will also have to lengthen the rising main to reach the cistern, as well as the distribution pipes from the cistern.

Booster pumps These incorporate an electric motor and must be wired into the power supply. There are two main types. A single pump is fitted between the mixer control and the spray and boosts the mixed supply to the spray. A dual pump is fitted to the supply pipes and boosts the hot and cold supplies separately before they reach the mixer. Depending on the model, a booster pump will provide sufficient pressure with as little as 150mm height difference between the water level in the cistern and the spray head.

Cistern-type electric heater
Household hot water supply pipe
Instantaneous shower unit
Shower supply direct from rising main
Waste pipe
Rising main

Pipework for an instantaneous shower Here, only a cold supply, direct from the rising main is needed. This is useful where there is no cold water storage cistern, as shown above with a combination hot water cylinder (see page 95 for details).

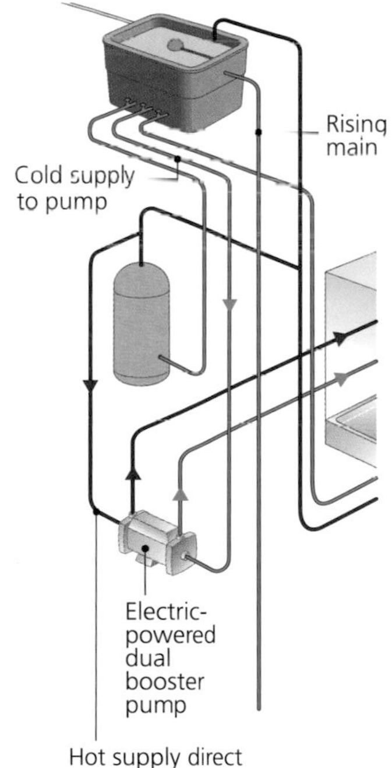

Boosting water pressure with a dual pump This boosts hot and cold water supplies separately. Some types of pump have a hot supply pipe direct from the cylinder casing, some from the vent pipe. A dual booster pump should be fitted by a plumber.

Choosing a shower

The type of shower that can be installed depends in part on your household water system (as described on page 83) and also on the strength of shower spray you like to have. But there are still choices to be made within each type.

Bath/shower mixer A shower spray combined with a bath mixer tap provides a shower for little more than the cost of the bath taps, and no extra plumbing is involved. The temperature is controlled through the bath taps, which may not be convenient, and will be affected by water being drawn off elsewhere in the home, unless you fit an (expensive) thermostatic type of bath/shower mixer. Pressure-balanced types are also available. No extra drainage required.

Power shower An all-in-one shower which incorporates a powerful electric pump that boosts the rate that hot and cold water are supplied to the shower head from the storage cistern and the hot water cylinder. A power shower is unsuitable where water is supplied from a combination boiler under mains pressure. Removing waste water from a power shower fast enough can be a problem. The shower tray must cope with around 20 litres a minute, so it may be worth fitting a 40mm (or even 50mm) waste pipe. Not normally suitable for use over a bath, unless it has a very good shower screen.

SAFETY WARNING

The hose to a showerhead must be fed through a retaining ring on the shower wall. This prevents the showerhead hanging in standing water in the bath or shower tray, avoiding potential contamination of the mains supply.

WETROOMS

A wetroom consists of a WC, basin and shower area. No shower tray or enclosure is fitted and water drains through a central drain set in a sloping floor, so the whole room must be waterproofed. This is not a DIY job. Wetrooms may have a powerful thermostatic mixer shower and body jets or a shower tower.

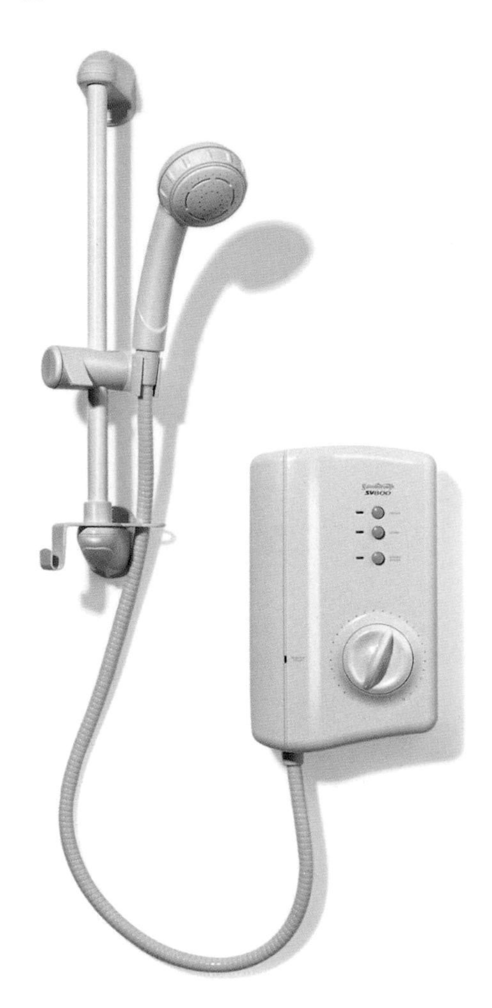

Instantaneous electric shower A wall unit plumbed in to a mains cold water supply, and heated by an electric element. The controls allow either less water at a higher temperature or more at a lower temperature, so the spray is weaker in winter when mains water is colder.

The unit must be wired to an electric power supply meeting Wiring Regulations requirements. This type of shower can be installed with any plumbing system and wattages up to 10.8kW are available.

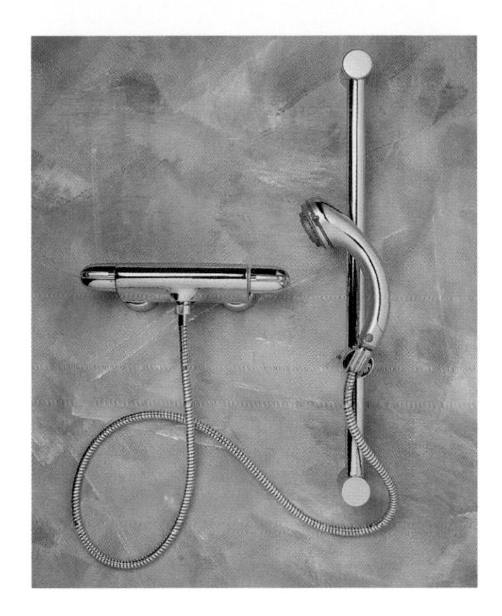

Manual and thermostatic mixers (above) These are wall-mounted units with hot and cold water supplies linked to a single valve. In a manual mixer, temperature and volume are controlled by one dial or separately. Thermostatic mixers are more expensive: their temperature control has a built-in stabiliser so water cannot run too hot or too cold. Computerised models have a control panel to programme temperature and flow rates and can store the data for each user. Pressure-balanced models are available for use when either hot or cold supply is at mains pressure; cistern-fed mixers can be linked to a pump to improve performance.

SHOWER FITTINGS

Spray roses Showerheads may be fixed or part of a handset on a flexible hose. The simplest have a single spray; multi-spray showerheads offer a choice of spray patterns selected by rotating the outer ring on the rose. Large diameter single spray showerheads offering a rain-style shower are also available.

Shower trays GRP-reinforced acrylic trays are light to handle and not easily damaged. A reconstituted stone or resin shower tray is heavy, stable and durable, but the floor must be level before it is installed. Shower trays come in sizes from 700mm square and are usually 110–185mm high; low level 35mm trays are available for 'walk-in' showers. Quarter circle and pentangle trays help to save space.

Shower tower or shower panel (below) A wall unit that incorporates a thermostatic mixer shower with a number of adjustable body jets. Tower units also have a fixed showerhead and a hand-held spray, and may be designed to fit into a corner or on a flat wall. Some can be installed over a bath while others are made for cubicles or wet rooms. A pump is usually needed to boost water pressure.

Installing a mixer shower

Most types of shower can be fitted either over a bath or in a cubicle. The fixings and pipe routes vary according to shower type, bathroom layout and shower location, but the installation method is basically the same.

Before you start Decide on the type of shower you want, bearing in mind that whatever showerhead you choose must either be fitted so as to prevent it coming into contact with water in the bath or shower tray, or it must have double-check valves fitted in the hot and cold pipes leading to the shower.

1 Mark the required positions of the spray head and shower control.

2 Plan the pipework to the shower control and how the waste water will be routed to the drainage system.

3 Fit the shower control. Some shower mixers are available as either surface-mounted or recessed fittings, and come with fixings and instructions. When fitting a recessed mixer, if possible mount it on a removable panel flush with the wall so that you have easy access to the controls.

4 Cut off the water supply and fit the water supply pipes. You can recess the pipes into the wall and then replaster or tile over them. However, they must be protected with a waterproof covering and have service valves fitted.

5 Fit the shower head and spray. For a separate cubicle, fit the base tray and waste fittings.

6 Connect the supply pipes to the shower control. An adaptor with a female screw thread (copper to iron) may be needed.

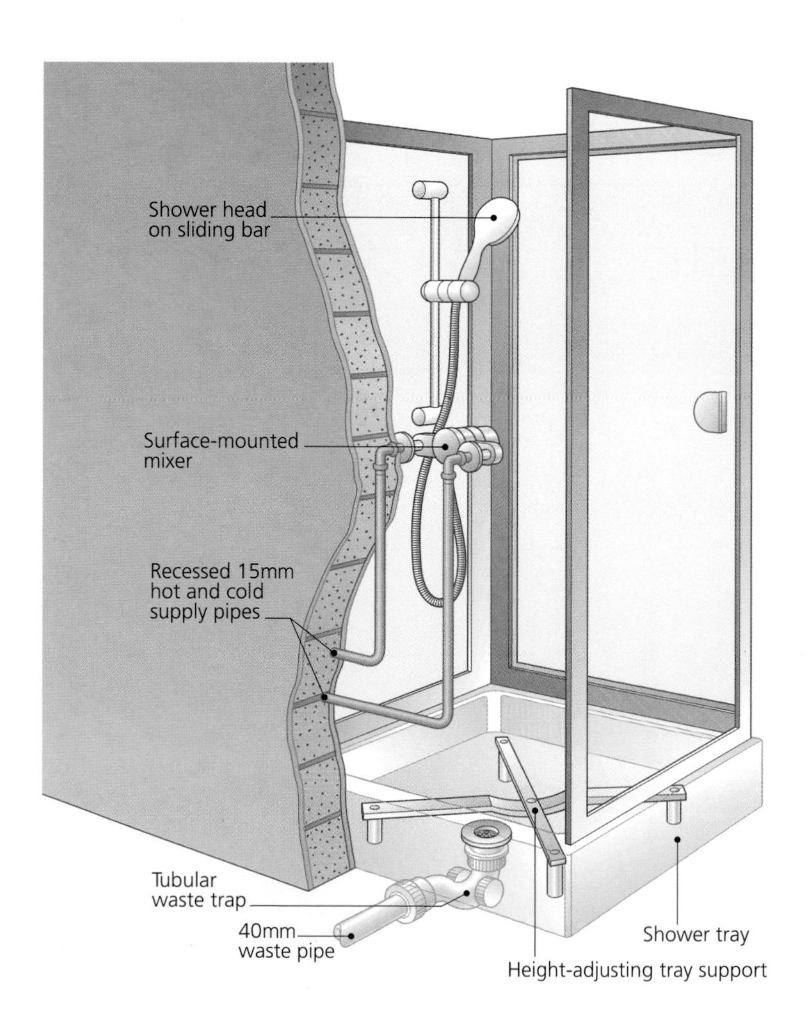

A typical cubicle installation
Screens are usually about 1.8m high. Panel widths can usually be adjusted by 25–50mm to allow for walls that are out of true. Doors may be hinged, folded (with panels shaped to keep water in), sliding with corner entry, or pivoted to give a wide entry without taking a lot of opening space. Some shower trays have an adjustable support by which the height can be altered so that the waste pipe and trap can be positioned either above or below the floorboards.

7 Restore the water supply and check the piping for leaks. Tighten any joints as necessary.

8 Fit screening panels and seal the joints between the wall and screening, and the tray.

INSTALLING AN ELECTRIC SHOWER

An electric shower needs only one plumbing connection – a branch pipe leading from the rising main (similar to that needed for a garden tap, see page 91), but this may be some distance away.

It will, however, need its own RCD-protected heavy-duty electric circuit run in at least 6mm² cable. The work need to be notified to the local authority building control department – unless you use a Part P-registered electrician to do the job. All plumbing work must be completed before the wiring connections are made.

Replacing a sink

You have several options if you plan to replace an old sink. A range of styles, sizes and bowl configurations is now available.

Some sinks have two tap holes, or holes for a two-hole mixer. Others have only one hole for a monobloc mixer. All have an outlet hole – either a 38mm hole for fitting a standard waste outlet, or a 89mm hole for fitting a waste disposal unit.There are two basic sink designs: lay-on and inset. Lay-on sinks rest on a base unit of the same size as the sink rim. They have been largely replaced by inset types, which fit into a cut-out in the worktop.

Sinks are available in a range of sizes and patterns. They are typically about 500mm wide and vary in length from about 780mm to 1.5m. Patterns vary from one bowl and a drainer in one unit, to combinations of one-and-a-half, two and two-and-a-half bowls with a drainer, two drainers, or no drainer at all. Drainers can be to the right or left side of the unit. Think about which will suit you best.

Stainless steel is the most common sink material. Enamelled pressed steel and more expensive polycarbonate are available in a range of colours. Easy-to-clean ceramic sinks are the toughest and most expensive, but china banged against them may chip.

Checking the waste outlet

Before choosing a sink, check the height of the waste outlet of the existing sink. The new sink may be higher or lower depending on bowl depth or the height of a new support unit. A typical waste-outlet height from a bowl 180mm deep in a unit 870mm high would be 690mm. If the new outlet will be higher than the old one, use a telescopic trap. If it will be lower, reposition the waste pipe.

If the pipe runs into an outside drain, you may need a new exit hole in the wall. If the pipe links to a single soil stack, make a new connection. The downward slope should be about 20mm for every 1m of run. Connecting the waste pipe to a single stack should be left to a plumber.

Fitting an inset sink

Most new sinks are designed for use with a monobloc mixer tap rather than two separate pillar taps. The sink may have one or two bowls.

The rim of the sink is sealed in its cut-out to stop water seeping beneath it, and is held in place by locking clips underneath. The monobloc mixer tap fits into a 35mm diameter hole in the sink unit, and has two flexible 10mm pipe tails projecting from its backnut. These are linked to the existing hot and cold water supplies with reducing joints. Fit an in-line service valve to each pipe before connecting it to the tap.

Tools *Two adjustable spanners for compression joints; long-nose pliers; screwdriver; jigsaw; power drill and twist drill bits; pencil. Possibly also pipe cutter.*

Materials *Inset sink (the correct fixings are usually provided); monobloc kitchen mixer with top and bottom washers; two compression joints (probably reducers); two in-line service valves; silicone sealant; PTFE tape.*

1 Mark the cut-out shape on the worktop according to the sink instructions. Check that the saw cut will not interfere with any structural parts of the base unit. Cut out with a jigsaw and seal cut edges with varnish.

2 With the sink on its face, seal round the rim and fit the earthing tag into the slot marked E on the rim.

3 Fit the securing clips provided to the sink, following the instructions. They are commonly hinged clips that are screwed to the sink rim and can be adjusted to fit worktops 27–43mm thick.

4 Fit the mixer tap, the waste outlet and any overflow pipes to the sink. Take care to position all sealing washers correctly.

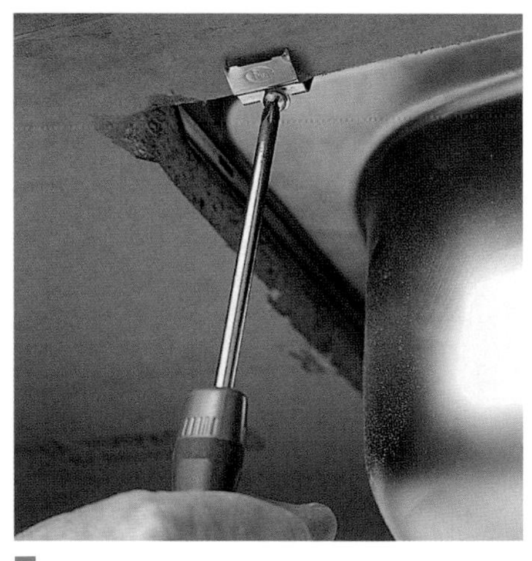

5 Fit the sink into the worktop, tightening the clips gradually in sequence.

6 Connect the trap to the waste outlet and add any overflow pipework. Then modify and connect the supply pipework.

CONNECTING A MONOBLOC MIXER TAP

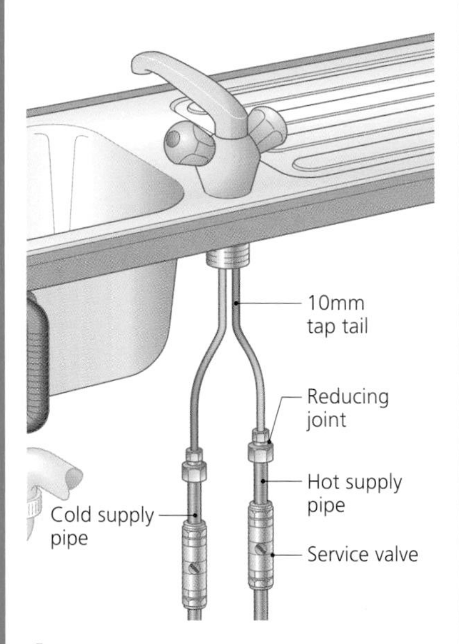

1 Check that the supply pipes are the right height for connecting to the tap tails. Cut the pipes if they are too high, and save the offcuts to link the new service valves to the tap tails. Label the pipes 'hot' and 'cold' so that you connect them correctly.

2 Fit a service valve to each supply pipe and turn it off with a screwdriver. Then add a short length of 15mm copper pipe to the outlet of each valve.

3 Connect the 15mm end of a reducing joint to each pipe tail.

4 Connect the narrower end of each reducing joint to its tap tail. You can bend the tap tails slightly to line up the connections, but take care not to kink them.

5 Screw up each capnut until it is finger-tight at both ends of each fitting. Hold each fitting steady with one spanner while you give each nut 1¼ turns with the other spanner.

6 Open each service valve and check each fitting for leaks. Tighten any weeping capnuts by another quarter turn.

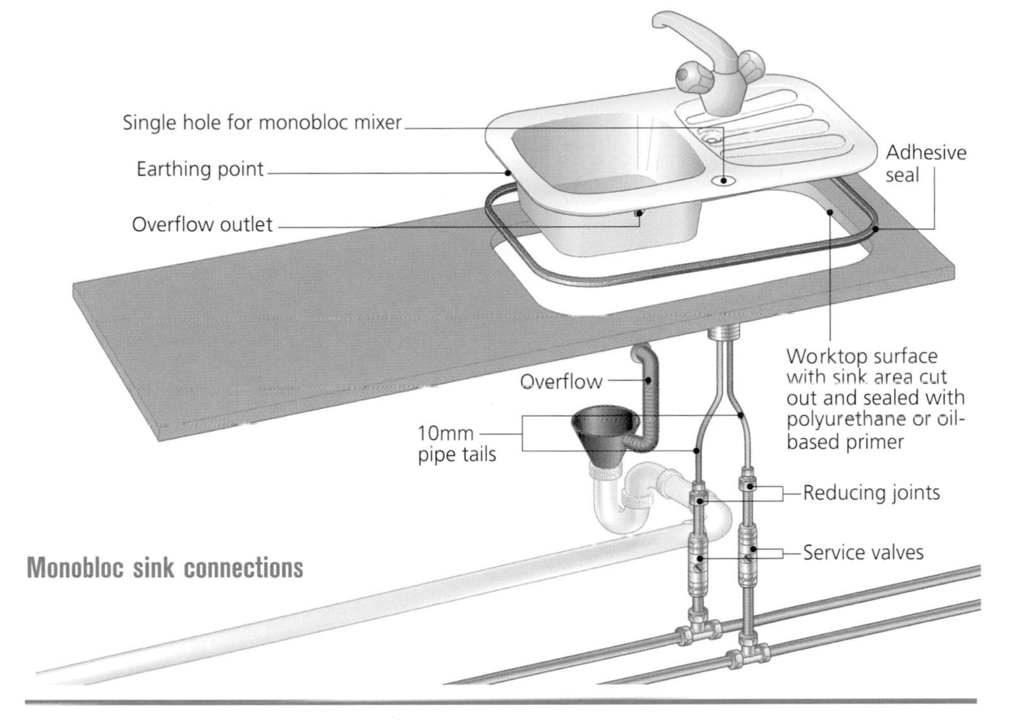

Monobloc sink connections

Plumbing in a washing machine or dishwasher

How you connect up the plumbing and drainage for a washing machine or dishwasher depends partly on how far away from the kitchen sink it is – the closer the easier.

Before you start Check the machine's instructions to see what the requirements are – especially for arranging the drainage. Some methods may not be allowed.

Tools *Hacksaw or pipe slice; half-round file; two adjustable spanners; medium-sized screwdriver; measuring tape; soft pencil; two spring-clip clothes pegs; spirit level. Possibly also shallow pan.*

Materials *15mm copper or plastic piping; one or two washing machine valves; one or two 15mm equal compression or push-fit tees; 15mm pipe clips; 40mm P trap, 40mm waste pipe and elbows, waste pipe clips.*

Connecting up the water

1 Turn off the water supply to the kitchen sink cold tap at the main stoptap.

2 It will often be easiest to connect directly to the rising main. Mark this at a convenient point (above the stoptap) with two pencil lines 20mm apart. Alternatively, remove the elbow on the pipe leading to the sink cold tap and replace it with a tee.

3 Cut through the pipe squarely with a hacksaw or pipe slice at the lowest point marked. A small amount of water will run out as you cut the pipe. Cut at the second mark on the pipe and remove the section of pipe. File the pipe ends smooth.

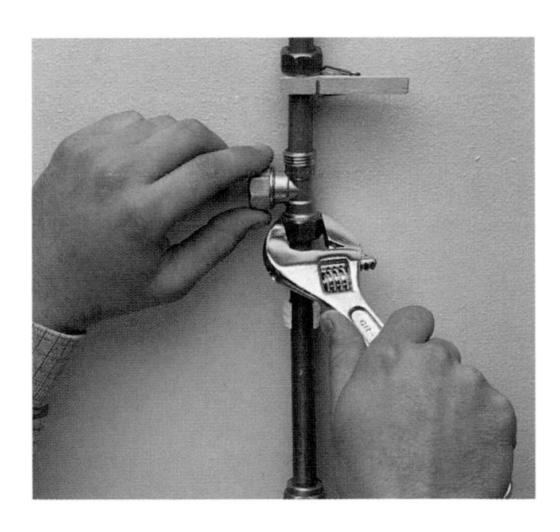

4 Use spring-clip clothes pegs to stop the caps and olives slipping down the pipe. Fit a tee joint to the pipe with the branch outlet pointing towards the machine.

5 Cut a length of 15mm pipe sufficient to get close to the machine and fit it into the tee. Use a pipe support for plastic pipes.

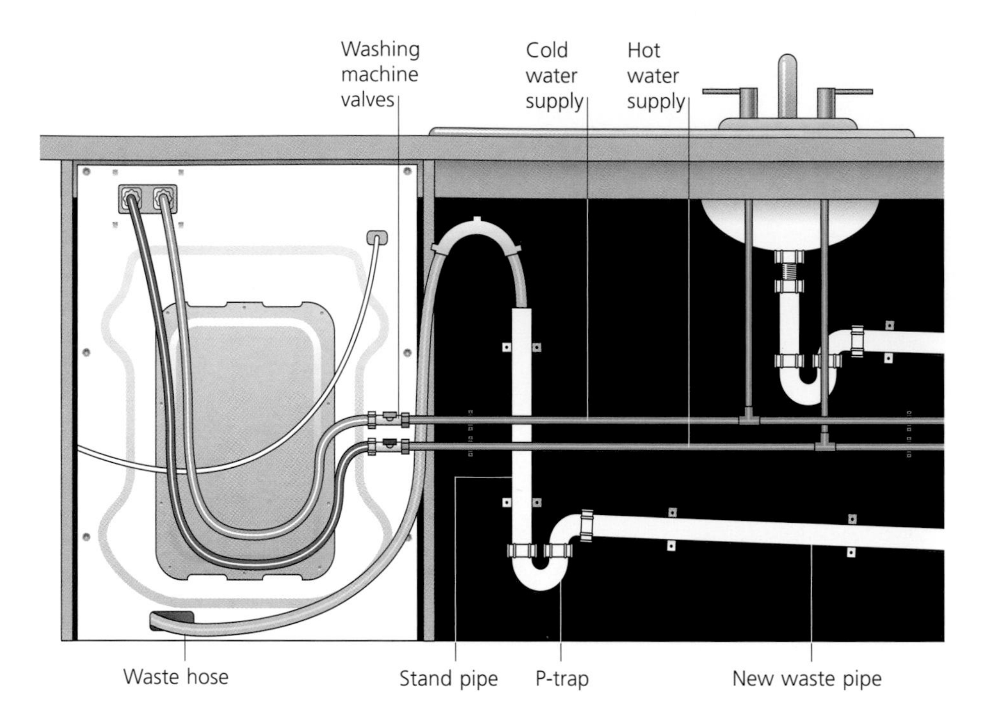

Washing machine connections The washing machine is shown connected to the hot and cold supply pipes under the kitchen sink via tee joints, with pipe leading to two washing machine valves. The waste hose is hooked into a stand pipe and new waste pipe is taken, via a P-trap, to the outside drains.

6 At the other end, fit the compression (or push-fit) joint end of a cold (blue) washing machine valve, fit pipe clips to support the pipe, and connect the cold-water hose.

7 For a washing machine, turn off the water supply to the hot tap over the kitchen sink (page 29). Cut the pipe and fit a tee joint, new pipe and (red) washing machine valve as above. Connect the machine's hot-water hose to the valve and restore the hot and cold water supplies.

Alternatively If the machine is close to the sink, you may be able to use self-cutting washing machine valves, which can be connected directly to the supply pipes.

Connecting the waste

1 Most washing machines require the use of a stand pipe into which the machine's hose is hooked. Install this, following the machine's instructions, on the wall close to the machine, securing it with clips.

2 Fit a P trap, followed by more 40mm waste pipe, leading to a convenient place on an outside wall through which it can be taken to the drains.

3 Make a hole through the wall and pass a length of waste pipe through it, connecting it with an elbow to the pipe from the trap on the inside and to the drains on the outside. Make good the hole in the wall.

Alternatively If the machine's instructions allow and the machine is positioned close enough, you may be able to connect directly to the nozzle of a 'washing machine' trap fitted under the kitchen sink.

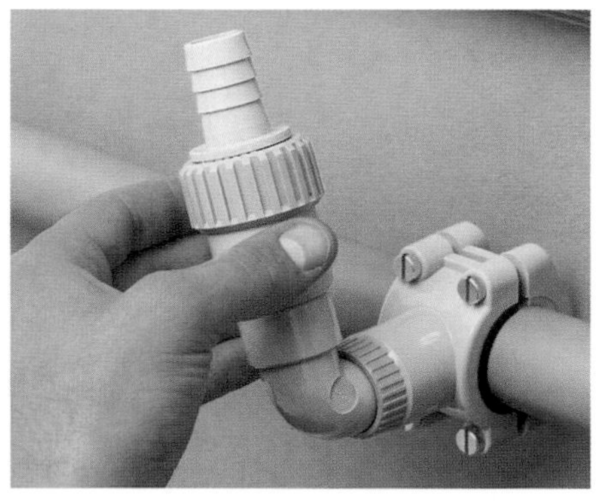

With some machines, you may be able to use a self-cutting washing machine drain kit which is simply secured to the waste pipe leading from the kitchen sink.

Installing an outside tap

Bib taps have a threaded nozzle which is suitable for fitting a garden hose. A tap installed against an outside wall should have an angled head; otherwise you will graze your knuckles when you turn the handle.

What you must do You do not need to inform your water company that you are fitting an outside tap – unless it is to be connected to some kind of mechanical or automatic watering system

You do, however, need to fit a double-check valve in the installation – either in the new pipe leading to the tap or built-in to the tap itself – to prevent possible contamination of the mains supply.

About the installation This job involves running a branch pipe from the rising main through the wall to the tap position.

The instructions given here are for fitting copper pipe run from a 15mm rising main with compression fittings; other pipe materials or fittings could be used.

- You should fit a stoptap into the branch pipe, as it allows you to do the job in two stages, and in winter you can cut off the water supply to the tap and drain it to prevent frost damage.
- As an alternative to making up the pipe run yourself, outside tap kits containing all the necessary parts are available from most DIY stores.
- The best way to make a hole through the brick wall of a house is with a heavy-duty power drill and masonry bit, both of which can be hired. Choose a bit at least 325mm long and 20mm diameter to allow the pipe to be passed through the wall easily.

Tools *Two adjustable spanners; hacksaw or pipe slice; pipe cutters; half-round file; power drill with masonry bits; screwdriver; two spring-clip clothes pegs; soft pencil; measuring tape; spirit level.*

Materials *Angled bib tap with threaded nozzle; 15mm stoptap; 15mm double-check valve; 15mm copper or plastic pipe; plastic pipe clips; 15mm equal tee; two 15mm elbows; wall-plate elbow; PTFE tape; wall filler; weatherproof pipe insulation.*

Positioning the tap

1 Mark the required position on the outside wall of the kitchen, as near as possible to the rising main.

2 Check that the mark is high enough for a bucket to be placed underneath the tap quite easily, and is at least 250mm above the damp-proof course in the house wall.

3 Make another mark for the hole through the wall about 150mm above the tap mark.

4 Take measurements from the hole mark to a point such as a window, so that you can locate and check the corresponding hole position inside.

5 Mark the position of the hole on the inside of the wall. Check that it will not interfere with any inside fitting and will be above the position of the main stoptap on the rising main.

Fitting the branch pipe inside

1 Turn off the main stoptap, then turn on the kitchen cold tap to drain the pipe.

2 If there is a drain valve above the stoptap, turn it on to drain the rising main, and prepare to collect the water that runs out in a container.

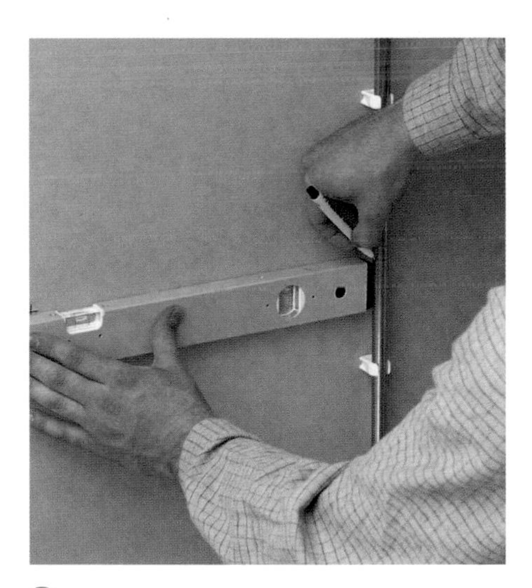

3 Mark the rising main at a point level with the hole mark on the inside wall. Make a second mark 20mm higher.

4 Cut through the rising main squarely

4 Cut through the rising main squarely with a hacksaw or pipe slice at the lower point marked. If there was not a drain valve above the main stoptap, be prepared for a small amount of water to run out as you cut through the pipe.

5 Cut the pipe at the second mark and remove the section of pipe. Use a file to smooth the pipe ends and remove burrs, and to square the ends if necessary.

6 Fit a capnut and olive over each cut pipe end. Use spring-clip clothes pegs to stop the nuts and olives slipping down the pipes. Fit the tee into the rising main with the branch outlet pointing towards the hole mark.

7 Cut a short length of pipe and connect it to the branch of the tee.

8 Connect the stoptap to the pipe, with its arrow mark pointing away from the rising main. Angle the stoptap so its handle leans away from the wall (see opposite page).

9 Close the new stoptap by turning its handle clockwise. You can now turn on the main stoptap and restore the water supply to the rest of the house.

10 Cut another short length of pipe and connect it to the outlet of the stoptap.

11 Connect the check valve to the pipe, making sure that its arrow mark points in the same direction as the stoptap arrow. Then complete the pipe run (see right).

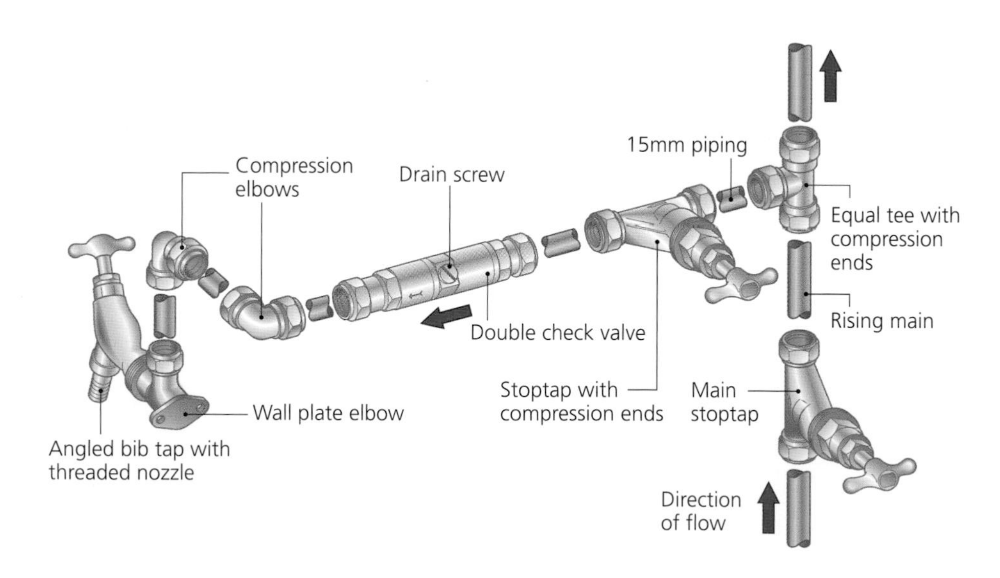

Pipework and fittings An outside tap is supplied by a branch pipe, commonly run from a tee joint fitted into the rising main. Instead of a tee joint, you can use a self-boring tap, which can be fitted to the pipe without turning off the water. No separate stoptap is then needed in the branch pipe.

Connecting the outside tap

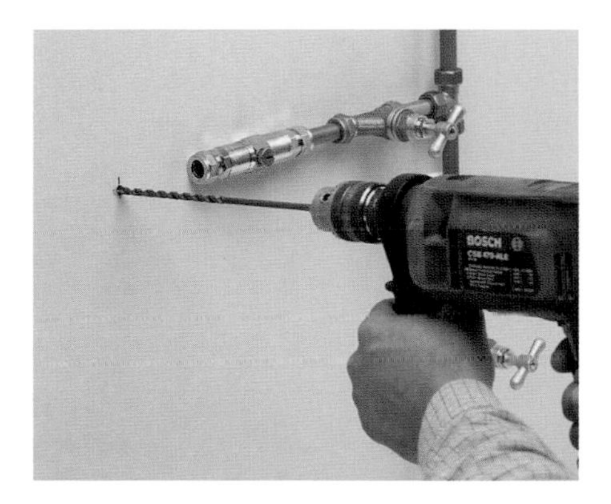

1 Use a long, slim masonry drill bit to drill through the wall from the inside first, making sure to keep the drill at right angles to the wall. Withdraw the bit at intervals to cool it and to pull out dust. Then use a 20mm masonry bit, working from both sides of the wall, to make a hole wide enough for the pipe to fit through easily.

2 Cut a length of 15mm copper (or plastic) pipe to fit through the wall with at least 50mm to spare.

3 With the pipe in position through the wall, slip on an elbow and measure the length of pipe needed to join the elbow to the double-check valve, allowing for the depth of pipe in each fitting.

4 Cut the pipe to the required length and smooth the cut ends.

5 Make the connections indoors.

6 Outside the house, cut the projecting pipe to leave only 25mm sticking out from the wall.

7 Fit another elbow to the projecting pipe, making sure that the free end of the elbow points towards the tap position mark.

8 Measure from the elbow to the tap mark and cut another length of pipe to fit the distance.

9 Fit the pipe to the inlet of the wall-plate elbow.

10 Fit the other end of the pipe temporarily into the elbow above, then hold the wall plate against the wall and mark the position of the screw holes.

11 Put aside the wall-plate elbow and pipe end, and drill and plug the wall.

12 Join the pipe to the projecting elbow and fix the wall-plate elbow to the wall.

Fitting the tap

1 Bind PTFE tape round the tail thread.

2 Screw the tap fully into the outlet of the wall-plate elbow.

3 If the tap is not upright when screwed home, take if off again, put one or two thin fibre washers over the inlet and refit. Keep on adjusting in this way until it is tight and upright.

4 Open up the new stoptap inside, and check all the pipe joints for leaks. Tighten if necessary.

5 Turn on the newly fitted outside tap and check that it is working properly.

6 Use wall filler or polyurethane foam filler inside and silicone sealant outside to seal round the pipe hole in the wall. Add weatherproof insulation to outside pipes.

Ways of saving heat

Money is wasted if water is heated and then not used. Inefficiencies in the plumbing system, or inefficient use of heaters, can also waste heat.

Insulate the hot water cylinder A 75mm thick lagging jacket on a hot water cylinder cuts down heat loss by about 70 per cent. A 140 litre cylinder without a jacket, maintained at a temperature of 60°C, loses enough heat every week to heat about 20 baths. All modern cylinders are foam-lagged by the manufacturer.

Keep hot water pipes short The length of pipe between the hot water cylinder and a hot tap is known as a dead leg, because hot water left in the pipe after each use of the tap cools and is wasted. The longer the pipe, the more the waste.

Water at 60°C travelling through a 15mm copper pipe loses heat equivalent to more than 1 unit of electricity for roughly each 300mm of run per week – enough to heat about 45 litres of water.

Where a hot water supply pipe to a basin or shower would involve a dead leg of piping of more than 6m long, it is wiser to use an instantaneous heater instead.

If you have an instantaneous water heater installed, position it as near as possible to the hot tap most often used – usually the one over the kitchen sink.

Avoid secondary hot-water circulation At one time, if a shower or tap was some distance from the hot water cylinder, there was a constant circulation of hot water to it by means of a return pipe back to the cylinder. This ensured that there was no delay in the arrival of hot water to the tap. Because of the heat lost, avoid such secondary circulation, particularly with electric heating.

Heat water only as needed A thermostat gives economical heating by controlling temperature, but even a well-lagged cylinder will lose heat (generally the equivalent of about 6 units of electricity a week). This can add considerably to costs if the heater is left on all the time. Switch on an immersion heater or boiler only about an hour before hot water is needed, and switch it off when it is not wanted. The most convenient way to do this is to have the heater fitted with a time switch that is set to turn it on for times of peak household use. Time switches have a manual override to allow use of the heater at other than the set times.

You can buy special immersion heater timers allowing several on/off periods a day (both 24-hour and 7-day version available). If you are on an electricity tariff offering cheap off-peak electricity, invest in an Economy 7 controller, which can be wired to control 2 immersion heaters bringing on one automatically at night and allowing the other one to be used during the day when required. Both timers and controllers are wired between the fused connection unit (or double-pole switch) on the immersion heater circuit and the immersion heater(s).

Install a shower Use a shower for daily cleansing and keep the bath for relaxed soaking. A bath takes about six times as much water as a normal shower.

Prevent scale formation in water pipes and appliances About 65 per cent of British homes – chiefly those in the south-east and Midlands – have hard or moderately hard water. The hardness is caused by a high concentration of dissolved calcium and magnesium salts, and is evident when, for example, soap does not dissolve properly and scale forms inside the kettle and round a tap nozzle.

Hard water drying on any surface leaves a crust of the salts behind, and at high temperatures the salts solidify into scale. When scale forms inside a domestic boiler or hot water cylinder, it insulates the water from the heat and wastes fuel, and pipes gradually become blocked.

Scale can be prevented or limited by a number of methods:

- Controlling the hot water temperature – scale starts to form above 60°C.

- Suspending scale-inhibiting chemicals in crystal form in the cold water cistern. They need changing every six months.

- More expensively, by plumbing a water softener into the rising main – beyond the kitchen tap and a branch to an outside tap. This leaves hard water, which most people prefer, for drinking.

- Adding a scale inhibitor on the (mains) cold feed to electric showers and combination boilers. This contains harmless polyphosphate crystals that are released into the water to prevent scale.

Ways of heating water

The average household uses 220–320 litres of hot water a day. Finding the best way of heating water for your home deserves some thought.

The commonest type of water-heating system is a hot water storage cylinder heated by a boiler (pages 122-123), probably combined with the central heating system. Various kinds of gas or electric heater can also be used to supplement the system, or as a complete system in themselves. They may be fed either by a low-pressure supply from the cold water cistern, or by a high-pressure supply direct from the mains.

Hot water cylinders

There are five main kinds of hot water cylinder to choose from – direct, indirect, combination, Economy 7, and unvented. All are sold pre-fitted with insulation.

- Direct cylinders are used where an immersion heater (see below) is the only way of heating the hot water – in an all-electric house or one with gas warm air central heating (or no central heating!). They have just two pipe connections – one for cold water coming in, one for hot water out – plus one or two immersion heater 'bosses'. Standard sizes are 450mm diameter and 900mm high (120 litre capacity) or 1050mm high (140 litre).
- Indirect cylinders are used where a boiler is used to heat the water (see page 16). It has two additional connections for the boiler pipes. Same sizes as direct cylinders.
- A combination cylinder is used where there is no main cold water storage cistern and cold taps are fed directly from the rising main – it has its own cold water cistern fitted above the main cylinder. Can be direct or indirect (may have two cisterns).
- Economy 7 cylinders are larger versions of normal cylinders (up to 210 litres) and are fitted with two horizontal immersion heaters – the lower one to heat the water at night, the upper one to top up during the day. Can be direct, indirect or combination.
- Unvented hot water cylinders are designed to take their cold water supply directly from the rising main. They come with various safety devices and must be installed by a qualified plumber. Can be direct or indirect.

Immersion heaters

Electric elements fitted into a standard storage cylinder to heat water. Sold in various sizes to suit different sizes of cylinder and horizontal/vertical application. All 3kW.

- There are three types: top-entry with one element extending almost to the cylinder bottom; top-entry with two elements – a long one for cheap night electricity and a short one for heating a small amount of water as needed; side-entry – usually a pair, one at the bottom to heat the entire cylinder and one at the top for heating small amounts.
- All types have thermostats. Can be fitted into a copper hot water cylinder as the sole means of heating, or as a supplement to a boiler. Can be renewed if the element burns out. Modern cylinders usually have 32mm or 57mm bosses in the dome or low in a side wall (or both) for heater fitting. Only water above the level of the element is heated. Expensive to use unless well insulated.

Instantaneous gas heater

The water is heated by gas as it flows through small-bore copper tubing. When the hot tap is turned on, the gas jets are ignited by a pilot light that burns continuously. The jets can be turned off.

- Large multipoint heaters can supply all household hot taps; smaller single-point types supply one tap only. The water supply is normally direct from the rising main. The heater can be fed from a cold water cistern if it is high enough – usually at least 2m above the highest tap – to give enough pressure.
- Useful where there is no cold water storage cistern. Only the water used is heated – there is no slow cooling of unused water. But the delivery rate is slower than from a cylinder, and the flow from one hot tap is interrupted if another is turned on. It is designed to raise water temperature by about 26°C, so in cold weather – when mains water can be near freezing – the heated water is either cool or slow running. In summer it may be too hot. Some have a winter/summer setting to vary the heat.
- The heater has a flue and must be fitted against an outside wall.

Instantaneous open-outlet electric heater

Small heater, supplied direct from the rising main, in which the water is heated as it passes through. Heaters with up to 10.8kW elements are designed for showers (page 84), those with a 3kW element are designed for washing hands.

- The water emerges through a spray nozzle. Useful for providing a shower where there is no suitable storage cistern supply or for hot water for washing hands in a cloakroom without hot supply pipes.
- These electric heaters usually raise the temperature by about 26°C, so water heat varies according to mains water temperature, but some types have a winter/summer setting. The flow may be interrupted when other taps are used, unless a compensating valve is fitted.

Replacing an immersion heater

An immersion heater element can burn out after long use or if it becomes coated with scale. Water that takes longer than usual to heat up may indicate scaling. Some immersion heaters are designed for use in hard water.

Before you start Although modern heaters have a thermostat and can be prevented from heating above 60°C – the temperature at which scale starts to form – they may be used as a supplement to a boiler that does not have the same degree of heat control.

To fit a replacement immersion heater to a combination cylinder (page 94), check the cylinder manufacturer's instructions. The heater may be on a plate assembly that can be withdrawn without draining the water.

Tools *Immersion heater spanner – box-type for deep lagging; electrician's screwdriver; adjustable spanner; hose clip.*

Materials *Immersion heater; PTFE tape. Possibly also 1.5mm² three-core heat-resisting flex; penetrating oil.*

1 If the cylinder is heated by a boiler, switch the boiler off. Then switch off the electricity supply at the consumer unit (fuse box).

2 Stop the water supply to the cylinder. Turn off the gatevalve on the supply pipe, if there is one, or drain the cold water cistern (page 29).

3 Turn on the bathroom hot taps and keep them running to draw off any water in the supply pipe.

4 Locate the cylinder drain valve. For an indirect cylinder (or a direct cylinder heated solely by an immersion heater) it is on the supply pipe where it runs into the base of the cistern.

5 Drain water from the cylinder as necessary – about 4.5 litres for a top-entry or high side-entry heater, or the whole cylinder for a low side-entry heater. Close the drain valve when you have finished.

6 Unscrew and pull away the immersion heater cover (below). Note which of the three conductors is connected to which terminal – use sticky labels as indicators if necessary – then disconnect them using an electrician's screwdriver.

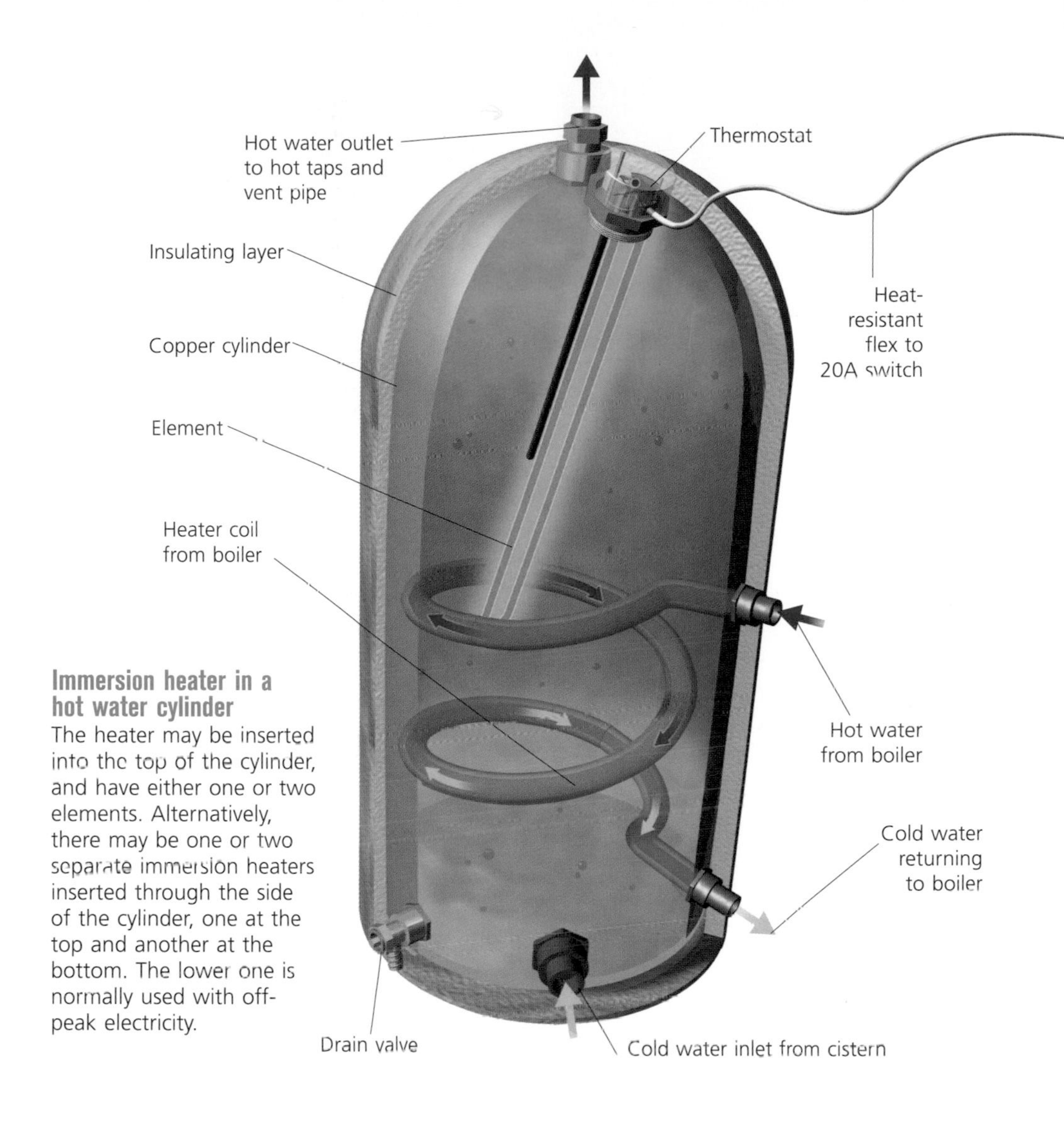

Immersion heater in a hot water cylinder
The heater may be inserted into the top of the cylinder, and have either one or two elements. Alternatively, there may be one or two separate immersion heaters inserted through the side of the cylinder, one at the top and another at the bottom. The lower one is normally used with off-peak electricity.

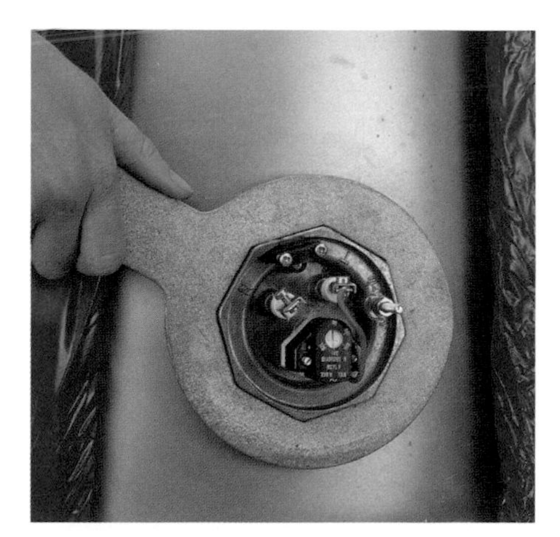

7 Use an immersion heater spanner to unscrew the old immersion heater and withdraw it from its boss. A flat spanner is suitable only if there is no deep lagging.

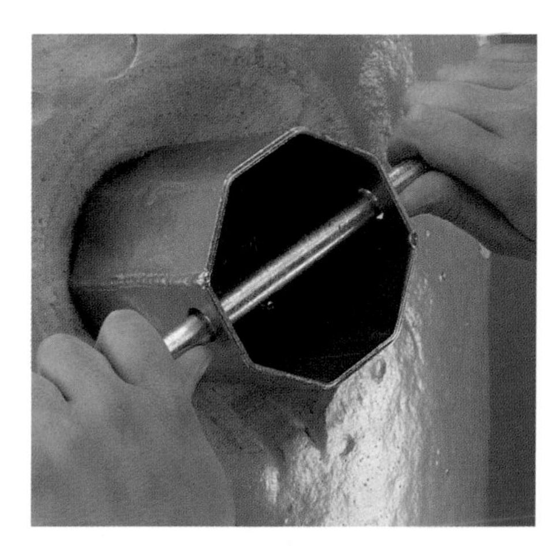

Alternatively For a deep-lagged cylinder, you will need a box-type immersion heater spanner, turned with a tommy bar.

8 Fit the heat-resistant fibre washer or rubber washer if supplied.

9 Fit the new immersion heater into the boss taking care to position it so it does not foul any internal heat exchanger coil.

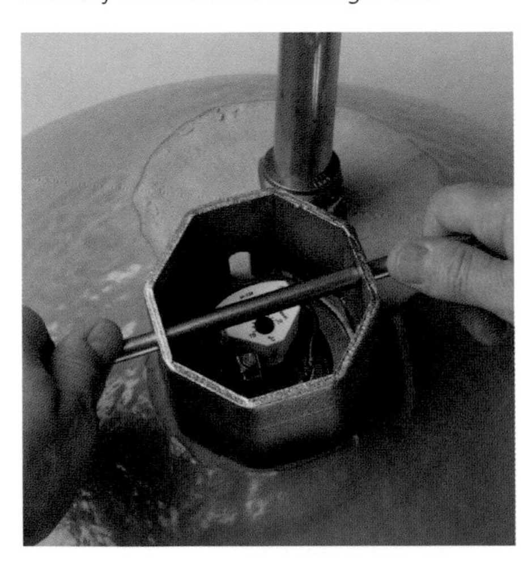

10 Tighten the heater into the boss. At this point you should fill the cylinder and check very carefully for signs of water seepage. If the mating surfaces are clean and the immersion heater is tight then you may need to tighten it a little more. Take care not to over tighten as you may damage the cylinder.

11 Re-connect the heat-resistant flex conductors to their terminals – normally brown (live) to the thermostat, blue (neutral) to the heater and earth (green/yellow) to the earth terminal.

12 Set the thermostat (below). Set a single thermostat (or the cheap-rate thermostat on an off-peak system) no higher than 60°C in a hard-water area (to prevent scale forming), or up to 65–70°C in a soft-water area. Where two thermostats are fitted, set the one for day-time operation to 50–55°C.

13 Replace the cover that protects the terminals and tighten the nut or screw that holds it in place. Do not cover the immersion heater to prevent heat loss because the cable will overheat.

SETTING THE THERMOSTAT

The temperature control can normally be adjusted with a screwdriver – the settings are marked round the screw. Some two-element heaters have one thermostat, some have two.

Choosing pipe insulation material

Insulating pipes is critical to prevent them from freezing and bursting in very cold weather, and to minimise heat lost as hot water travels around the system.

Self-adhesive foam wrap Thin foam insulating wrap, 50mm wide, is supplied in rolls usually 5m or 10m long. Some types have a metallic finish.

There is no formula for estimating how much wrap to buy – it depends on the size of the pipes and how large you make the overlaps. Buy and use one or two packs, then work out how much more you will need to complete the job.

Before you fix the lagging, make sure that the pipes are clean and dry. Peel off the backing paper and wind the material round the pipes. Overlap the tape as you wind, especially at bends. This flexible lagging is also useful for insulating awkward fittings, such as stoptaps.

Plastic foam tubes Easy-to-fit plastic foam tubes are split down one side and have to be eased open to fit them round the pipe. They are secured with adhesive tape wrapped round at intervals, or with purpose-made clips. Tubes are available to fit 15mm, 22mm and 28mm pipes. Plastic foam tube is slightly more expensive than self-adhesive foam wrap, but is easier to fit.

Foam tubes are available in two wall thicknesses. In most cases the standard grade is sufficient, but if you live in an area that often experiences severe frosts or if your pipes are particularly exposed, it is worth investing in the thicker material.

Glass fibre blanket Pipes that are boxed in can be insulated by stuffing glass fibre blanket around the pipes.

COLD-WEATHER CHECKS

- Make sure no tap is left dripping. If that is not possible, put a plug in the bath or basin overnight. Drips cause ice to block waste pipes.
- Never allow cisterns to overfill. Water in overflow pipes can freeze, causing the cistern to overspill.
- In a long cold spell, open the loft hatch occasionally, to let in warmth from the house.
- If you leave the house for short periods, keep the central heating switched on, but turned down to the minimum setting.
- For long periods, drain the plumbing system by closing the main stoptap and opening all the taps. When the water stops running, open the drain valve near the stoptap. For central heating, see page 116.

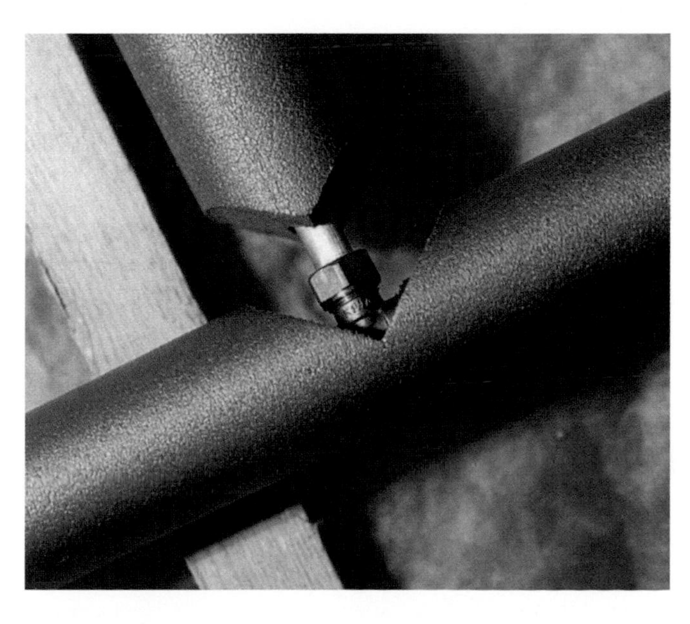

Cut neat joints Make 45° cuts in split-sleeve foam tubes with scissors or a sharp bread knife so you can form neat joins at elbows and tees. Use PVC insulating tape to keep the joints tightly closed and avoid a freeze-up.

Insulating hot and cold water pipes

Hot and cold water pipes that are exposed to the cold should be lagged to prevent winter freeze-ups.

Before you start Concentrate first on pipes that run across a loft, above an insulated floor, and those that run along outside walls in unheated rooms. Overflow and vent pipes that are exposed to the cold should also be lagged. Some pipes are boxed in. To lag them, unscrew the box and stuff pieces of glass fibre insulation all round the pipes. Make sure all pipes are clean and dry before you start.

Lagging pipes with self-adhesive foam wrap

Tools *Scissors.*

Materials *Rolls of self-adhesive foam wrap.*

Self-adhesive foam wrap is useful where there are many bends in the pipes and it would be difficult to use flexible foam tubes.

1 For pipes in the loft, begin work at the cistern. Cut pieces of foam wrap to a workable length with scissors.

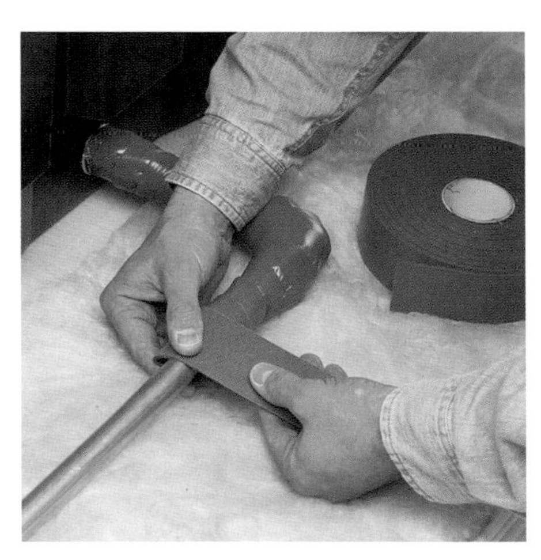

2 Wrap foam round the pipe, making generous overlaps of about one-third of the width of the wrap. Take care to cover the pipe well at bends – these are the vulnerable areas most likely to freeze.

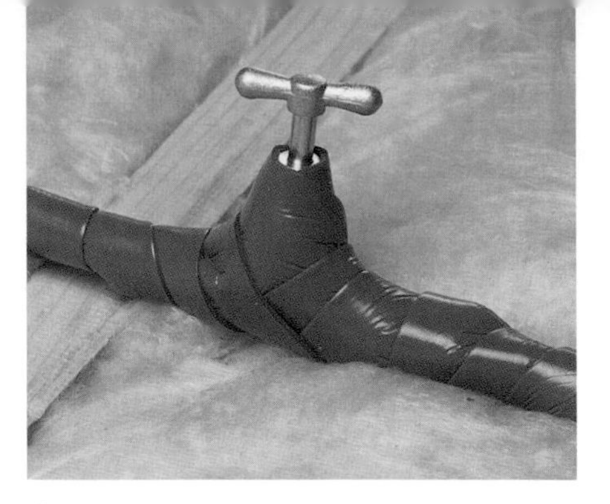

3 Take the wrap around any valves or stoptaps as you meet them, leaving only the handle exposed.

Lagging pipes with flexible foam tube

Tools *Scissors; serrated knife.*

Materials *Foam tube to match pipe size; adhesive tape; plastic clips.*

1 Lag the pipes leading from the cistern first, if you are insulating pipes in the loft. Wrap plastic adhesive tape around the first tube to hold it in place, even if the tube is one of the self-locking types. Push it up tight against the cistern so that the tank connector joint is covered.

2 Butt-join the tubes where they meet and wrap tape round the join to hold them tight. Cut the tube at 45° to fit it round elbows and tee fittings, and tape the joints. Alternatively secure joints with plastic clips.

3 Cut the tube to fit round the body of a gatevalve as closely as possible – or use self-adhesive foam wrap for these.

Lagging a cold water cistern

Never lay lagging under a cold water cistern. Heat rising through the ceiling will help to prevent a freeze-up.

Before you start Purpose-made jackets are available to insulate most cisterns. Measure the cistern's diameter and height, if it is round, and its height, length and width if it is rectangular.

It does not matter if the jacket you buy is too large, since the sections can be overlapped. If the cistern is an odd size or shape, or you want to provide extra insulation, use plastic-sleeved glass fibre loft insulation blanket. This is easier to handle than unsleeved blanket and will not release fibres into the air as you handle it.

Using glass fibre blanket

Tools *Steel tape measure; scissors.*

Materials *150mm thick wrapped glass fibre blanket; string.*

1 Wrap the cistern in a continuous length of blanket, which you have cut with scissors so that the edges will meet. Tie a length of string round the blanket.

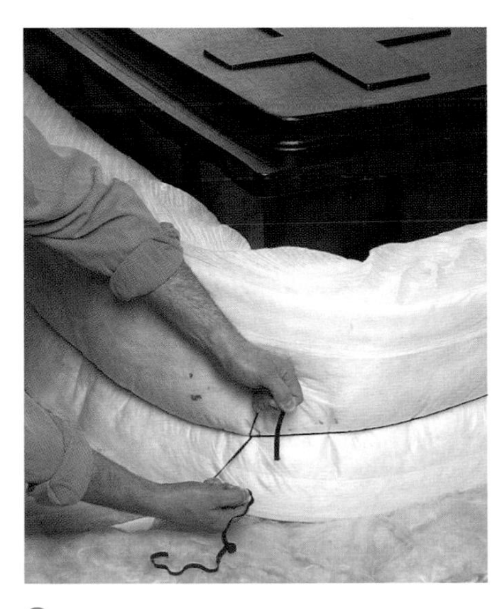

2 If necessary, wrap a second length of blanket round the cistern. Cut it to length and tie it on the same way as the first layer.

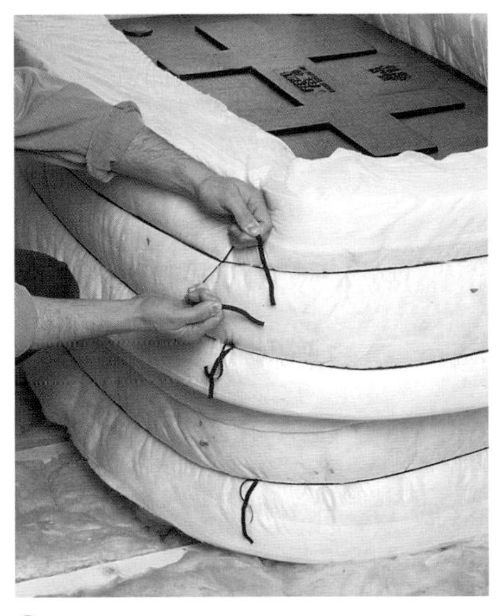

3 Extend the top layer beyond the top of the cistern, to create a small rim to hold the cistern's lid in place.

4 Measure the size of the cistern lid. Cut a length of blanket to match and staple the ends closed. If the blanket is too wide, squeeze it up tightly to fit. Do not cut the blanket along its length.

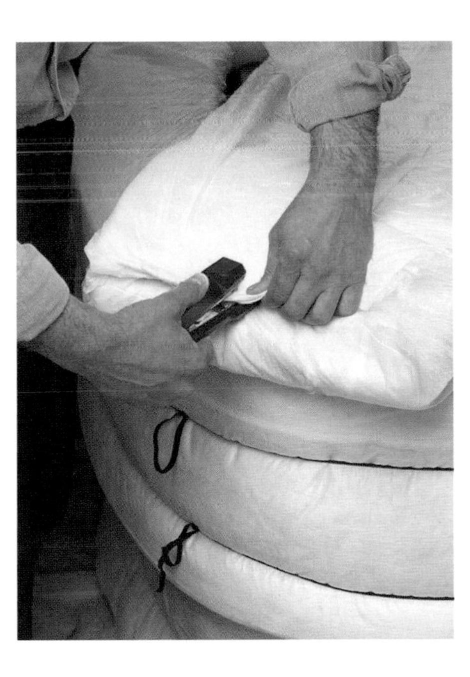

Insulating a high-level cistern

High-level cisterns in roof spaces are more at risk from freezing than cisterns sitting on the loft floor. This is because the heat rising from the house is not trapped under the cistern.

For the same reason, the pipes leading up to the cisterns are also at risk. If these pipes freeze, then there is a chance that the central heating boiler could explode. The most vulnerable pipe to freezing and bursting is the cold mains supply to the central heating header tank, because the water in this pipe rarely moves so it has plenty of time to freeze.

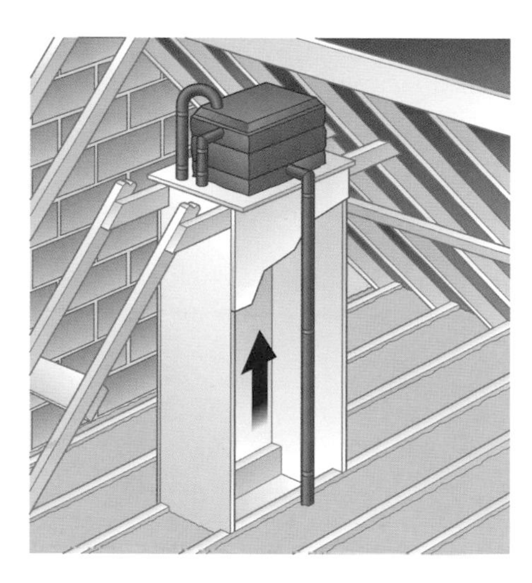

The best way to solve this problem is to build an enclosure directly under the cistern in order to funnel heat upwards from the house and give added protection to the pipes. You can do this from any insulation material but the easiest to work with is fire-retardant polystyrene. It is lightweight, easy to cut and self-supporting. You can buy sheets of it very cheaply from builders' merchants. You may find even cheaper sheets with corners broken off.

Tools *Fine-toothed saw, hacksaw blade or serrated bread knife; tape measure; plumbline or spirit level.*

Materials *Sheets of polystyrene; polystyrene tile adhesive; 100mm wire nails or wooden meat skewers; adhesive tape.*

1 Use a plumbline to drop a vertical position from the edges of the high level cistern down to the ceiling below. Cut and peel back the loft insulation at this point so that you have a clear area of ceiling corresponding to the shape of the cylinder.

2 Lay a piece of polythene over the ceiling at this point, cut to the same size as the base of the enclosure. This is to stop vapour from the house, which evaporates through the ceiling, from entering the new enclosure and condensing on the underside of the cistern.

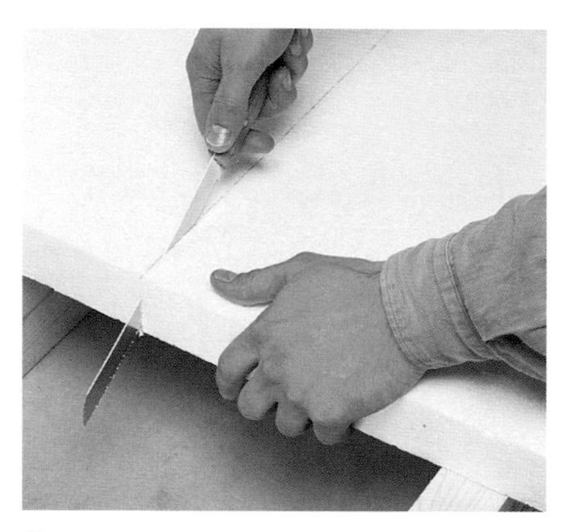

3 Using the saw blade or knife, cut the polystyrene sheet to form walls from the ceiling to the cistern sides. Stick the edges together with a small amount of tile adhesive and push the nails or skewers through the corners to hold the sheets in place while the adhesive dries.

4 Use a little adhesive on the ceiling around the polystyrene to seal any gaps and hold the sheets in place. Remember that lofts can be quite draughty in high winds and polystyrene can be blown around if it is not fixed.

5 Bring the polystyrene right up to the underside of the cistern. The cistern must be supported on an independent structure. Tape the top of the polystyrene to the sides of the cistern support.

6 The cistern or cisterns must be very well insulated. A thin jacket is not sufficient if the cistern is near the slope of the roof. Use glass fibre blanket over the top of existing insulation to increase protection.

7 Make sure that the vent pipe, which goes over the cistern, has an open passage to the cistern to discharge any water. The vent pipe outlet must be above the water line.

Controlling your central heating

Efficient temperature and time controls can save a great deal of money on fuel bills.

Room thermostat This temperature-sensitive switch is set to a pre-selected room temperature. It sends an electrical signal to switch the heating on when the air temperature falls below the pre-set level, and off when it rises above the level.

On fully pumped systems the room and hot water cylinder thermostats operate motorised valves. When these are opened they in turn switch on the boiler and pump.

On gravity hot water systems the room thermostat operates the pump and the boiler is switched on and off by the programmer.

A room thermostat is best placed in a draught-free spot on an inside wall away from direct sunlight, about 1.5m above floor level, and away from any heat sources.

Thermostatic radiator valve The best means of controlling the temperature in each room is to fit a thermostatic radiator valve (TRV) to each radiator. The valve opens and closes according to the temperature in the room. If the room is cold, a full flow is allowed through to the radiator. Then as the room warms up, the valve closes to reduce the hot water flow through the radiator.

Rooms facing south and rooms with open fires or other heat producing appliances, such as an oven, benefit most from TRVs.

Most systems are suitable for use with thermostatic radiator valves. Seek expert advice on which ones to buy.

Leave one or two radiators without TRVs to act as a bypass, in order to maintain open circulation in the system. Alternatively, a bypass pipe can be installed just after the pump. The best type of bypass is a pressure-operated valve which opens progressively as the TRVs close the radiators down. It also helps to cut surging noises.

TRVs do not control the central heating pump and boiler, so they must be combined with a room thermostat or a boiler energy manager.

Programmers Time controls range from simple switches to complex electronic programmers.

The most useful can time room heating and domestic hot water separately, so water heating can be turned

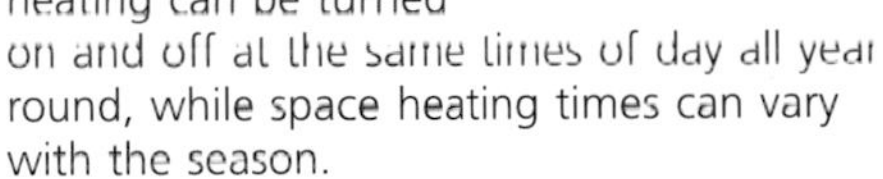

on and off at the same times of day all year round, while space heating times can vary with the season.

Electronic types can give you three control periods a day and different settings for every day of the week. Some even have a 'holiday' setting.

Water heating control The hot water temperature is often controlled only by the boiler thermostat. So hot water to the taps is at the same temperature as the water supplied to the radiators. This is probably hotter than necessary.

An electric thermostat fitted on the outside of the hot water cylinder will restrict the temperature of the water inside. It switches a motorised valve on and off to control the flow of water passing through the heating coil inside the cylinder.

Boiler energy management Sophisticated devices make sure that the boiler works only when needed.

A boiler energy manager will reduce wasteful short cycling on a boiler – that is, when 'hot water only' is selected on a conventional central heating programmer, the boiler will continually switch on and off to keep the water in the boiler at the selected temperature. It will do this even though the hot water cylinder is already full of hot water. This 'short cycling' can add as much as 30 per cent to fuel bills.

The boiler energy manager will also take account of outside temperatures, and will regulate the central heating system accordingly. For example it will override the setting and delay the start time of the central heating on warmer days.

Updating your central heating programmer

An old-style programmer can be easily replaced with a more modern one.

Before you start All new programmers now have an industry standard backplate. This allows you to swap one for another with more settings such as 'weekend' settings and 'one hour boost'. If you have an old-style programmer you will need to rewire the backplate to suit the new programmer. Full instructions will be supplied with the new model.

Tools *Electrician's screwdriver.*

Materials *New programmer. Possibly also pencil; paper; masking tape.*

1 Turn off the power supply to the heating system and remove the fuse.

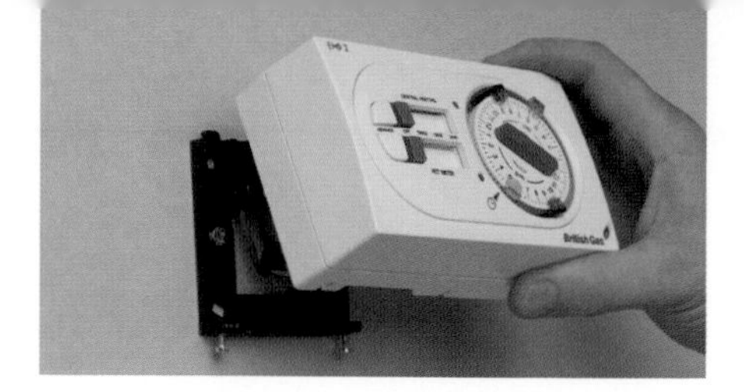

2 An old-style mechanical programmer with very limited time settings and options can be removed from the backplate by undoing the retaining screws.

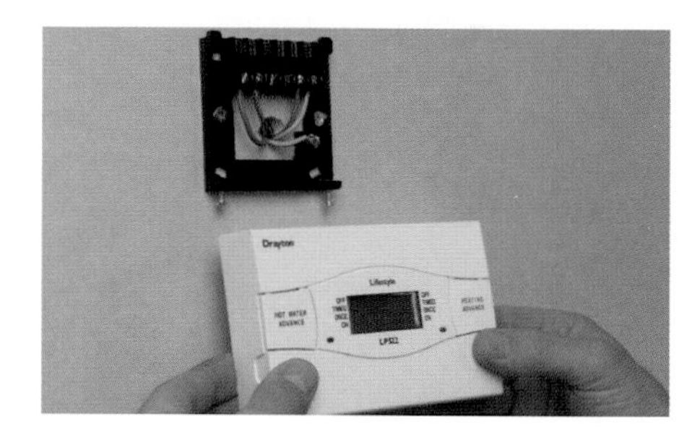

3 If your backplate looks like the example shown above, you can fit a new programmer directly to it. You may need to change the settings on the switch at the back to suit your heating system.

An older style gravity hot water system which only allows the heating to operate if the hot water is switched on is known as a ten position programme. If you can turn the heating on without the hot water then you can leave the slide switch set to 16 positions.

4 The new programmer must be tightened onto the backplate so the pushfit terminals are secure. You can then switch on the power and set the on-off times.

5 If the new faceplate does not fit the old backplate, study the wiring carefully and label each wire clearly. Sketch the old connections, then disconnect the cable cores from their terminals.

6 Remove the old backplate. Attach the new one, using the manufacturer's instructions and your sketch to wire it up.

7 Push the new faceplate into position and restore the power to the wiring centre.

Central heating problems: what to do

If the hot water or heating stop working, there are some simple but useful checks that could save you a call-out fee, or help you to give a plumber helpful information about the nature of the problem.

No central heating or hot water

• Check that the programmer is set to 'on'. It may have been turned off in error.

• Check that the thermostats are turned up to the correct level.

• Check that the electricity supply is switched on and that the fuse has not blown. If the power is on and the fuse is working, but the programmer is not receiving power, there may be a loose wiring connection. Call an electrician to check the wiring and trace the fault.

• If a motorised valve is fitted, check that it is working properly. Slide the manual lever to open the valve. If there is resistance, the valve is not opening. This could indicate a burnt-out motor. Call a central heating engineer.

• If the pump is not working, you can try to start it manually. Turn off the central heating system and wait until the pump is cold. Remove the screw in the middle of the pump and turn the impeller (the pump's manual starter). On some models this is a small screw that is turned with a screwdriver, on others there is a small handle attached.

• If this does not work, try tapping the pump casing sharply, but gently, with a mallet two or three times.

• If this does not work, remove the pump (page **114**), and flush water through it with a hosepipe. Do not submerge it in water.

• If this does not work, replace the pump (page 114).

• If the pump is running, but the boiler does not light, check that the pilot light is on and that the gas supply is turned on at the meter. If you have an oil boiler, check that the fuel is turned on and that there is oil in the tank. Check that the filter is clear.

• If the pilot light is not lit, follow the procedure in the handbook or on the boiler casing to relight it. If the flame will not stay lit, the flame failure device probably needs renewing. Call a central heating engineer.

• If a combination boiler will not light, check on the pressure gauge that the water pressure is at least 1 bar. If it is below 2.5 bar, top it up via the mains filling point. If it is above 2.5 bar, call a central heating engineer.

• If the mains pressure to the house as a whole has dropped (check by running the taps), call your water supply company for advice.

The central heating is working but there is no hot water

• Make sure that the thermostat on the hot water cylinder is set to 60°C.

• Check that the motorised valve (if fitted) to the cylinder is open (see No central heating or hot water, left).

• Bleed the air-release valve beside the hot water cylinder (if there is one). The valve is usually located on the pipe which enters the heating coil.

Hot water working but all radiators cold

• Make sure the room thermostat has not been altered – it should be set to 21°C.

• Check that the motorised valve is opening properly (see No central heating or hot water, left).

• Check that the pump is working – it should be warm to the touch and you should be able to feel if it is running. If not, try to restart it (see No central heating or hot water, left).

Noise in a central heating system

Unusual noises in a central heating system should not be ignored. The cause may be something quite easy to rectify.

Creaking in the floors and walls

Pipes expand as they heat up and contract as they cool. If the pipe is gripped tight by timber or a wall, or if it is in contact with another pipe, a creaking noise will occur when it gets hot or cold.

1 Pack some felt or pipe lagging around the pipes where they come up through the floorboards.

2 If that does not work, take up the floorboards around the source of the noise.

3 If one or two pipes are lying in a notch in a joist, and there is no room for movement, make the notch slightly wider by cutting down with a tenon saw and chiselling away the waste. Do not make the notch deeper; you may weaken the joist. Ease a piece of felt or pipe lagging under and between the pipes.

4 With the floorboards up, use a rasp to enlarge the holes through which the pipes rise to the radiator, if they are tight. Cover the pipes with pipe lagging where they pass through the boards.

5 Where pipes run the same way as the joists, make sure they do not sag or touch. Hold up sagging pipes with pipe clips fitted on struts between the joists. Put lagging between any pipes that touch.

6 If the pipes go through a wall, sleeve them with fire-resistant material, such as glass fibre matting, or pack it in around them, tamping it fairly hard with a screwdriver.

Boiler noise

- Loud banging noises or sounds like a kettle boiling coming from a boiler indicate the presence of corrosion in the heating system and possible scale build up within the boiler's heat exchanger. Both can be removed by adding a central heating system cleaner to the feed-and-expansion cistern (or injecting it into a radiator for a sealed system – page 117) and then running the system hot for at least an hour. After this, drain the system and flush through with clean water until it runs clear.
- To prevent future corrosion and scale formation, add a corrosion inhibitor (system protector) to the final water fill.
- Noise may occur if the water flow through the boiler is insufficient. With modern lightweight gas boilers, water flow rate is particularly important.

Banging in the pipes

- Banging noises in the pipes may be due to overheating. To find the cause of the fault, start by checking that the boiler thermostat is working properly. Turn the boiler off but leave the pump running to help to cool the system down. Then turn the boiler on and turn up its thermostat. If you do not hear a click, turn everything off again and call a central heating engineer.

• You can cut down on the amount of noise transmitted along copper pipework by cutting out a section and inserting one or two plastic push-fit fittings into the run. If the noise persists, replacing troublesome sections of copper pipework with semi-flexible plastic pipe may be the answer.

Sound of rushing water in the pipes

Air that has entered the system, or gas that has formed as a result of internal corrosion, can cause a noise in central heating pipes like the sound of rushing water.

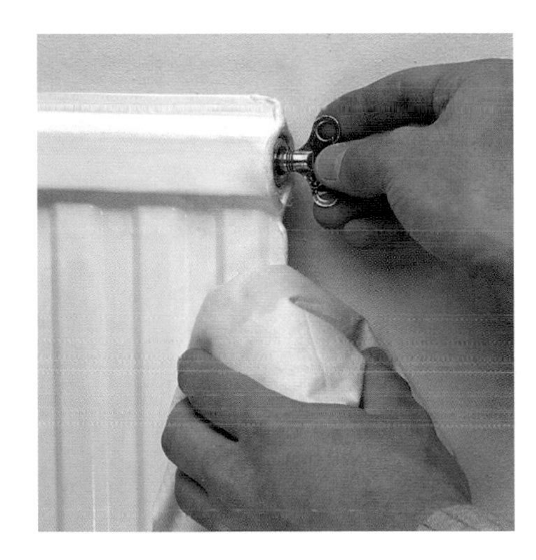

Try releasing the air from the air vents on the radiators, and any other venting points in the system. If the noise continues, it may be a symptom of serious faults that could eventually damage the whole system. Poor positioning of the safety open-vent pipe (see pages 18-19) could be the cause. Get expert help.

Humming in the pipes

An annoying humming sound usually comes from the pump. Call in an expert to find the cause.

• Anti-vibration pump brackets can be fitted that may help to reduce the problem.

• Pipes may vibrate if they are too small for the amount of water they have to carry.

• The pump speed may be set too high. Try turning down the speed control knob on the pump body by one setting. If this fails to cut the noise and also makes the radiators take longer to heat up, call in a central heating engineer. He may suggest relocating the pump.

Balancing a radiator circuit

Hot water is carried from the boiler to the radiators by a flow pipe, which branches off to supply each radiator. Cool water leaves each radiator at the opposite end and joins a return pipe carrying it back to the boiler.

As water flows round the radiator circuit, it cools down, so the radiators nearest to the boiler receive hotter water than those furthest away. This is dealt with by adjusting the water flow through each radiator, so that the less-hot ones get more water through them. Adjustment is carried out using the lockshield valve on the radiator: the one with a plain cover.

Tools *Two clip-on radiator thermometers; spanner; small screwdriver.*

Materials *Sticky labels; pencil.*

1 Two or three hours before you intend to start work on the radiators, turn off the central heating system to allow the water in the radiators to cool down.

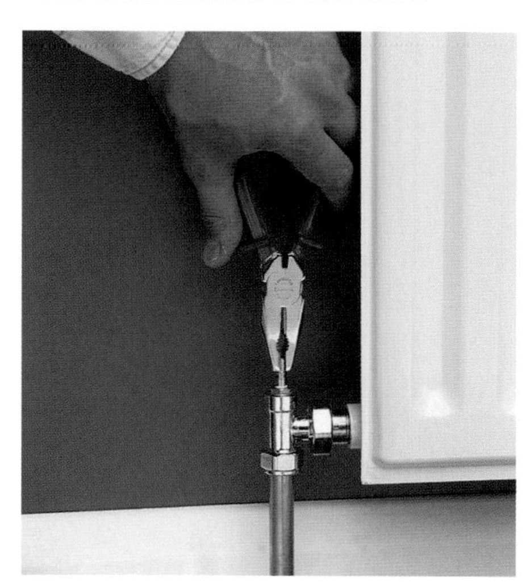

2 Open all lockshield valves and handwheel valves fully.

3 Turn on the central heating system. Work out the order in which the radiators heat up, and label them accordingly.

4 Clip a radiator thermometer onto the flow pipe bringing water into the first radiator, and one onto the return pipe.

5 Turn down the lockshield valve until it is closed, then open it slightly. Adjust the flow until the temperature of the flow pipe is roughly 11°C higher than that of the return pipe.

6 Repeat for all radiators in the circuit, working in the order as labelled. The lockshield valve on the last radiator will probably need to be fully open.

A typical radiator circuit

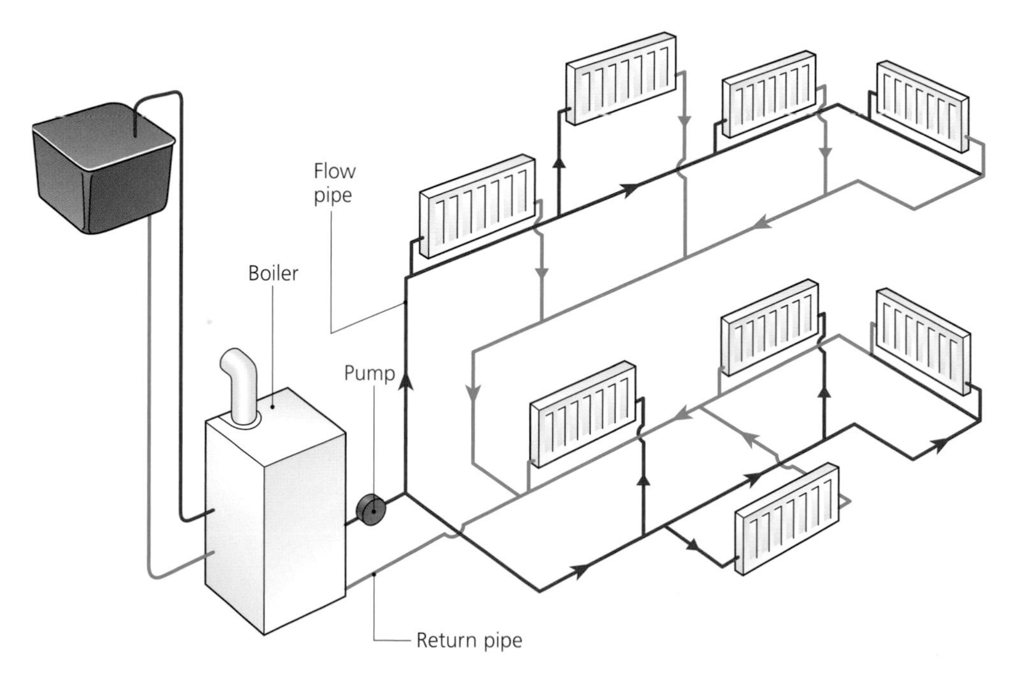

Leaks in a central heating system

Never ignore leaks in a central heating system. Fresh water that is drawn in to replace the lost water contains free oxygen which can cause radiators and cast iron boilers to rust and scale to form.

Before you start Internal leak sealants similar to the radiator 'weld' used in cars can be used to seal very minor leaks. Pour the sealant in through the feed-and-expansion cistern. Do not use leak sealant in a sealed system.

A leaking pipe joint

Most leaking pipe joints are compression fittings, which can be tightened with a spanner. Tighten the joint slightly, no more than a quarter turn. If this does not stop the leak, do not tighten any further as this will damage the joint.

1 Drain the system to below the level of the leak. Undo the nut on the leaking joint and pull the pipe out slightly.

2 Wrap two or three turns of PTFE tape around the face of the olive where it meets the joint. Tighten the nut.

3 If the leaking joint is soldered, drain the system. Heat the joint with a blowtorch and take it apart, then replace it (see page 57).

A leaking radiator valve

If the leak is from the compression joint below the valve, drain down the system to below the joint. Then call a plumber or repair the joint yourself. Use PTFE tape to cure a leak from the union nut connecting the valve to the radiator.

1 The most common cause of leaks on radiator valves is where the PTFE sealant tape has run up the shiny chrome as the valve is screwed into the radiator. This leaves no sealant on the thread. Use a hacksaw blade to 'break in' the thread by striking it across as if you were striking a match. The small barbs will hold the tape in place. Some threads come ready broken.

2 The valve seal to the tail is made at the olive and not around the thread, as many believe. Improve the seal by wrapping some PTFE tape around the olive face.

3 When you tighten or loosen a radiator valve always counter the force with another spanner to prevent any strain on the pipe.

Thermostatic valves

1 Modern thermostatic valves (TRVs) use a 15mm compression fitting on each port and are bi-directional so you can fit them on the flow or the return.

2 TRVs have closing off caps to allow you to remove the radiator safely. Never rely on the thermostatic valve closing the water off. They can open suddenly as the temperature drops in the night.

3 Remove the closing off cap and fit the thermostatic head. To enable you to screw it down fully, the head must be screwed onto the body with the setting on 'maximum'.

A leaking valve tail

The leak may be from the valve tail screwed into the radiator. Use a radiator spanner to remove it. Cover the male thread on the valve tail with PTFE tape and replace the tail.

A leaking radiator vent

If the radiator air vent leaks, drain the system to below the vent. Remove the air-vent fitting using a radiator spanner. Bind the screw joint with PTFE tape, and replace the fitting.

A leaking radiator

A small jet of water from the body of the radiator is called a pinhole leak. It is caused by internal corrosion and can happen within a few weeks of the system being fitted if the debris that collects during installation has not been removed, or if air is being drawn in.

Turn off the valves at each end to relieve the pressure. Then remove the radiator and leave the rest of the system running. Before fitting a new radiator, flush out and clean the system using a non-acidic cleaner.

Repacking a radiator gland

If a radiator valve weeps water from under the cap, the packing gland is worn. You can replace the packing with PTFE tape or thread-sealing fibre, sold by plumbers' merchants. Some radiator valves cannot be repacked; instead they have renewable O-rings which can be replaced with a kit.

Tools *Small adjustable spanner; small screwdriver; PTFE tape; silicone grease.*

1 Turn off the valve. If it continues to leak, close the lockshield valve at the other end of the radiator.

2 Remove the cap from the leaking valve and undo the small gland nut. Slide it up out of the way.

3 Pull a length of PTFE tape into a string and wrap this around the spindle four or five times.

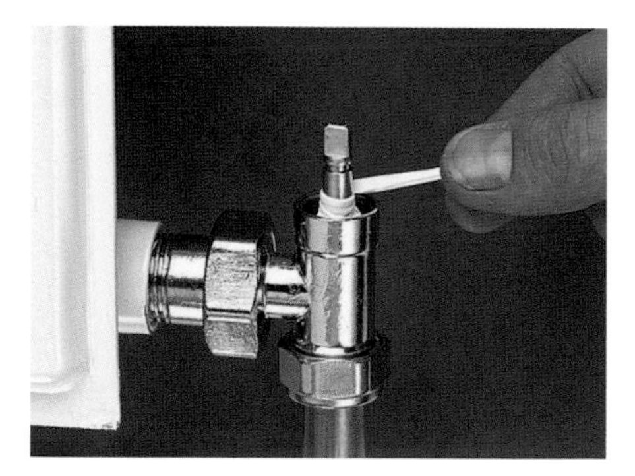

4 Use a small screwdriver to push the tape down into the valve body. Smear on silicone grease and re-tighten the gland nut. Replace the head and turn the valve back on.

Replacing a radiator valve

Although radiator valves normally last for years, the time will come when it is necessary to replace one.

Before you start Drain the heating system (page 31). As the system drains, open the vents on the upstairs radiators and then those downstairs.

Tools *Two adjustable spanners. Perhaps hexagonal radiator spanner.*

Materials *New radiator valve; wire wool; PTFE tape.*

1 Undo the nut that connects the valve to the radiator by turning it counter-clockwise. To stop the valve rotating, hold the body of the valve upright with a second spanner.

2 Undo the capnut that connects the pipework to the body of the valve, by turning it clockwise (as seen from above). Lift the valve away and let the capnut slip down the pipe.

3 Separate the new valve from its tail piece and check whether the valve will fit the old tail piece. If not, use the radiator spanner to remove the old tail and clean the threads in the radiator with wire wool.

4 Fit the new tail piece to the radiator if necessary after first wrapping PTFE tape clockwise around its threads to make a watertight seal. Tighten it with the radiator spanner. Check that the valve lines up with both the tail piece and the pipework.

5 Place the valve over the pipe, slide up the cap nut and screw it up finger-tight. Finger tighten the nut securing the valve to the tail piece before tightening both nuts with a spanner – use a second spanner to brace the valve when tightening the nuts.

6 Re-fill the system and bleed each radiator to get rid of trapped air. Close the radiator air vents one by one and check the valve for leaks – tighten nuts if necessary.

Removing a radiator

It may be necessary to remove a radiator in order to flush out sludge that has built up inside, replace it or decorate behind it. This can be done without draining the whole system.

Tools *Polythene sheets; old towels; rags; two bowls; pliers; two large adjustable spanners; absorbent paper; hammer; hexagonal radiator spanner.*

Materials *PTFE tape. For replacement: new radiator the same size as the old one; new radiator air vent; radiator plug.*

1 Lay a polythene sheet and old towels on the floor around the radiator. This could be messy.

2 Shut the control valve by hand. Then remove the cover from the lockshield valve and use pliers or a small spanner to shut it too. Count the number of turns that this takes and write it down.

3 Put a bowl under the control valve and disconnect the union nut. Take care not to distort the pipe. Water will flow out (there may be a lot, so have bowls ready).

4 Open the air vent to increase the flow of water.

5 When it has stopped, undo the union nut on the lockshield valve. Some more water may come out.

6 Block the open ends of the radiator with twists of absorbent paper.

7 Lift the radiator off its brackets and carry it outside. You may need help.

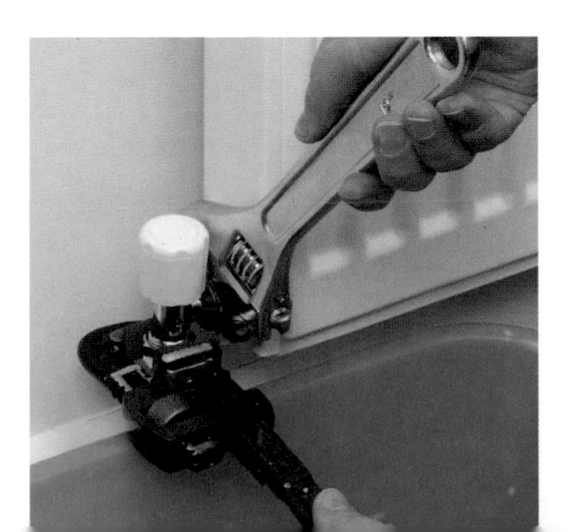

Fitting a replacement

1 If you are replacing a radiator, but keeping the valves, remove the valve tail pieces from the old radiator. Turn the valve tail counter-clockwise (when looking at the end of the radiator).

2 Hold the new radiator in position to check if the wall brackets need repositioning.

3 Wind PTFE tape round the thread of the valve tail pieces. Screw the tail pieces in place.

4 Fit a new air vent at the same end of the radiator as before, using PTFE tape as for the valve tail. Use the radiator spanner to tighten it in. Fit a new plug if there is an open tapping in the other top end.

5 Lift the radiator onto the wall brackets and reconnect the valve union nuts.

6 Open the valves to fill the radiator with water. Let air out through the air vent, and check for leaks. Reset the lockshield valve to its original position.

WARNING

Turn a TRV down to zero (or fit the special screw-down cap provided) before disconnecting the tail pieces and removing the radiator. Otherwise there is a risk of flooding if the temperature drops and the valve opens.

Relocating a radiator

Changing the position of a radiator can free up valuable wall space when you change the layout of a room.

Before you start Drain the heating system (page 31). As it drains, open the vents on the upstairs, then downstairs radiators. Remove the radiator leaving valves in place.

With the floorboards raised, cut back the pipes leading to the radiator from the main flow and return pipes: either fit a stop-end to each pipe below floor level or fit a blanking plug to the compression tee fitting. Extend the central heating pipework so that you will be able to have two new pipes coming up through the floor at the new position, but don't fit the final new pipes yet.

Tools *Tape measure; pencil; power drill; masonry drill bit; screwdriver; spirit level; hacksaw or pipe cutter; spanners.*

Materials *Radiator; wall mounting brackets; two radiator valves; 50mm No. 12 screws; wall plugs; 15mm copper pipe; compression plumbing fittings; PTFE tape.*

1 Before removing the old brackets from the wall, measure their spacing and distance above the floor and use these measurements to mark the bracket positions at the new location. If the wall is a stud wall, make sure the brackets will be secured into studs – or fit horizontal battens across the studs to take the bracket screws.

2 Drill holes for wallplugs in a solid wall (pilot holes on a stud wall) and screw the brackets to the wall or to your two new battens (use new screws if necessary).

3 Use a spirit level to check that each bracket is vertical and that they are aligned horizontally – if anything, there should be a very slight rise towards the end where the air vent is fitted. The brackets are designed to allow adjustment.

4 Hang the radiator on the brackets, check that it is level, and fit the new pipes from the main flow and return so that they align with the two radiator valves. You may find plastic pipe easier to use – see page 55.

5 Connect the pipes to the valves, re-fill the system and bleed each radiator to get rid of trapped air. Close the radiator air vents one by one and check for leaks at all connections at the old and new radiator locations – tighten nuts if necessary.

Boxing in pipes

Exposed pipework can be concealed in boxes. Insulate hot water and heating pipes before covering them.

Tools *Bradawl; screwdriver; pencil; drill bit with twist, masonry and countersink bits; panel pins; hammer; nail punch; spirit level; plane.*

Materials *Timber battens; one-piece joint blocks (often used to assemble flat-pack furniture); screws and wall plugs; hardboard or 3mm thick MDF; foam pipe insulation.*

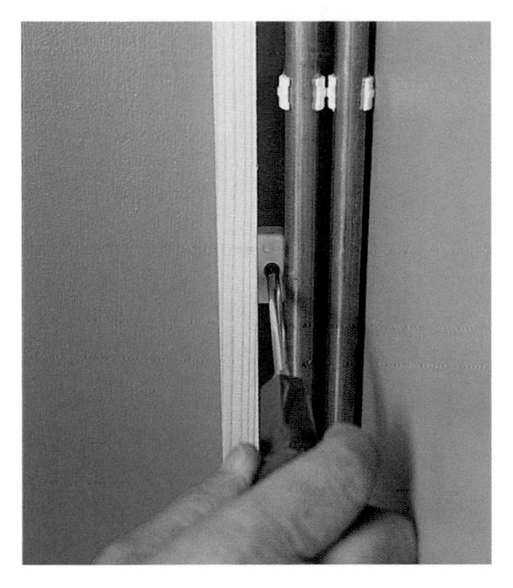

1 To box in a group of pipes in a corner, fix a batten a little wider than the depth of the pipes to the wall, using one-piece plastic joint blocks at 1m intervals.

Fix a second batten on the other side of the pipes if they are not in a corner.

2 Insulate pipes if necessary. Cut a strip of hardboard or MDF wide enough to cover the batten and the pipes and pin it to the batten with panel pins or screw it home.

3 Where pipes run along a skirting board, fit a horizontal batten above the pipes, then fit a slimmer one at floor level, and finish off with a cover strip, ready for painting or papering.

Radiators that do not heat up correctly

If your radiators are not giving out enough heat, check them all and make a note of which are cool and which (if any) are hot. Some may be cooler at the top or bottom.

Radiator cool at the top

Air is trapped at the top of the radiator. Turn off the central heating. Then use a radiator bleed key to open the air vent at one end of the radiator. Air should start to hiss out. When water appears, close the vent. Hold a rag under the vent to catch the water escaping from it. Turn the heating on again.

If radiators need bleeding more than once a year, air is entering the system and this can cause corrosion. There may be a serious fault that needs expert attention. Some systems have one or more extra bleed points on the pipes either upstairs or in the loft. Manual bleed points are opened with a screwdriver.

On an automatic air valve the small, red plastic cap must be loose in order for air to escape. If it is tight, unscrew it.

Radiator cool at the bottom and hot at the top

Sludge (black iron oxide) produced by internal corrosion can build up at the bottom of a radiator and stop circulation. Remove the radiator, take it outside and flush it through with a hose. If more than one radiator is affected, drain and flush the whole system before adding a system cleaner to the feed-and-expansion cistern as described on page 106. After final flushing, add a corrosion inhibitor to the final water fill.

Top-floor radiators cold

Cold radiators upstairs only, often indicate that the feed-and-expansion cistern is empty. The ballvalve may be faulty (page 48).

Refill the feed-and-expansion cistern so that there is just enough water to float the ball when the water in the system is cold. The extra space accommodates expansion of the water in the system as it heats up. After re-filling the system, bleed all the top-floor radiators (page 113).

Top-floor radiators hot, lower radiators cold

This is almost certainly due to pump failure (see Changing a central heating pump, right).

Cold radiators throughout the house

Most likely to be wrongly-set room thermostat, a faulty motorised valve or a faulty pump – see *Hot water working but radiators all cold* on page 105.

Radiators farthest from the boiler are cool

The system is not properly balanced (page 107).

Top radiators heat up when hot water only is selected on programmer

Hot water naturally rises above cooler water. On a gravity driven system, hot water for the hot water cylinder is prevented from creeping into upstairs radiators when the heating is switched off by a mechanical valve, called the gravity-check valve. It is situated on the flow pipe to the upstairs radiators.

If the gravity-check valve is stuck in the open position, the pipe on either side of the valve will be warm. Call a central heating engineer to replace it.

Changing a central heating pump

You can change a central heating pump without first draining down the whole central heating system, provided that there are service valves fitted on each side of the pump.

Before you start Domestic pumps are now a standard size, but if the old pump was longer than the new one, you may need adapters to fill the gaps.

When you go to buy a replacement from a plumbers' merchant take all the details of the old pump with you. Measure the length of the old pump, and the diameter and type of the connections. Most domestic pumps have 1½in BSP threaded connections.

Also make a note of the type of pump and the setting of its output regulator (domestic pumps are available with different ratings).

Tools *Electrician's screwdriver; bowl; towels; pipe wrench or adjustable spanner; pencil and paper.*

Materials *New pump.*

---

1 Switch off the electricity supply to the central heating system at the fused connection unit – or at the consumer unit.

2 Make a note and sketch of how the electrical wiring on the old pump is connected. It may be helpful to label each conductor. Then disconnect the conductors with a screwdriver.

3 Close down the service valves on each side of the pump using the valve handle or an adjustable spanner. If there are no isolating valves, drain down the system.

4 Put a bowl and towels under the pump ready to catch any water that escapes when you remove it.

5 Unscrew the union nuts holding the pump in place. Turn them counter-clockwise (facing along the pipe towards the pump). Remove the old pump.

6 Fit the new pump in position with the new sealing washers in the unions to prevent leaks.

7 Open the isolating valves (or refill the system) and check that the unions are watertight.

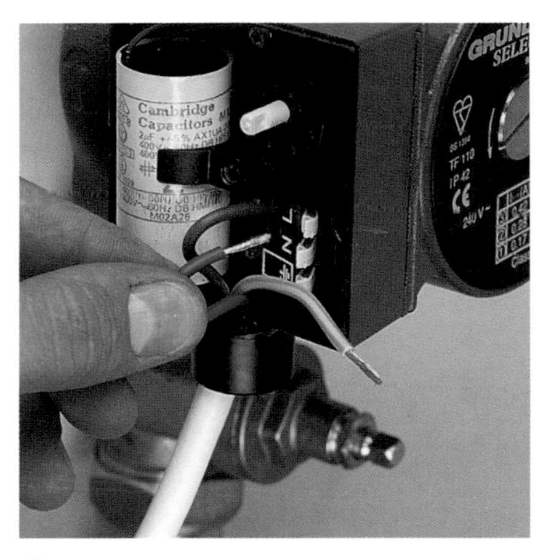

8 Dry the pump carefully to remove any traces of moisture; reconnect the wiring.

9 Test the newly installed pump by switching on the electricity supply and turning on the central heating system at the programmer or time switch. You may also need to turn up the room thermostat to get the system going.

10 Once the central heating system has started up, check that the safety open-vent pipe over the feed-and-expansion cistern does not discharge water when the pump starts or stops. If it does discharge water, seek expert advice.

11 If you have had to add much fresh water to the cistern, bleed any air out of the system in order to guard against future corrosion and to protect the new pump.

HELPFUL TIPS

The central heating pump is designed to run full of clean water. Any air or sludge that gets into the water can damage the pump, so keeping the system full of water and limiting corrosion are essential. Never run the pump unless it is full of water.

If your pump is out of use in summer, as is usually the case with a combined gravity and pump-driven system, run it for a minute once a month to keep its impeller free.

Repairing a motorised valve

If a motorised valve ceases to open, its electric motor may have failed.

Before you start Use an electronic tester (see below) to check whether the valve is receiving power. If it is, you will need a new motor. You should be able to buy one from a plumbers' merchant. There is no need to drain the system, but you must switch off the electricity supply to the central heating system. Just turning off the programmer is not enough, because a motorised valve has a permanent live feed.

1 Take off the valve cover and undo the retaining screw that holds the motor in place. Push the lever to open the valve, and lift out the motor. Cut off the connectors to disconnect the two motor wires.

2 Insert the new motor, then let the manual lever spring back to a closed position. Fit the retaining screw and tighten. Strip the ends and connect the wires, using the two connectors supplied with the new motor. Put back the cover.

3 Check the new motor by turning on the power and running the system.

Preventing a freeze-up

If you turn off your central heating while you go on a winter holiday, there is a danger that the system will freeze and a pipe will burst.

Lagging only reduces the speed of heat loss, so eventually the temperature of an unused system will drop to the level of the surrounding air. With a gas-fired or oil-fired system, leave the heating on and turn the room thermostat down to its minimum setting if you will only be away a few days.

Using a frost thermostat

For a long holiday, you could have a frost thermostat installed (it is also called a low-limit thermostat). It overrides the controls and turns on the system when the temperature approaches freezing point. Rising air temperature makes it turn the system off again.

Adding antifreeze

You can also add antifreeze to the water in the central heating system (but **not** the water in the main cold water cistern). Tie up the ball valve arm in the feed-and-expansion cistern and pour in antifreeze according to the maker's instructions. Then drain off enough water via a drain valve for the antifreeze to be drawn into the system. After restoring the water level in the cistern to the correct level, turn the central heating on for a few minutes in order to thoroughly mix the antifreeze with the water.

ELECTRONIC TESTERS

Electronic testers works in one of three ways: metal detection (pipes and cables), density detection (studs and joists) and voltage detection (live electricity). You will find detectors with one, two or all three of the functions. Note that metal detectors may not work with foil-backed plasterboard.

Protecting a system against corrosion

The life and efficiency of a central heating system can be increased by adding a corrosion and scale inhibitor.

Before you start Test the water in the system every year to see how corrosive it is. To do this, drain a sample of the heating system water into a jar and place two bright (not galvanised) wire nails in the jar. Screw the lid on. Wait for a week.

If the nails rust and the water turns a rusty orange, the water is seriously corrosive (or has no corrosion inhibitor) and you must take immediate action. If the water remains fairly clear and the nails do not rust, then no further action is necessary, since the water in the system has lost its free oxygen. A few black deposits are acceptable.

Finding out where air is entering the system

The most common cause of corrosion is air in the system. If the radiators need bleeding more than once or twice a year then too much air is being drawn into the system and this must be eliminated.

The most common areas where air gets into the system are a leaking joint on the suction side of the pump, or through the feed-and-expansion cistern.
Leaks around pumps can be repaired in the same way as other leaking joints. But if air is entering through the feed-and-expansion cistern you will need expert help.

You can find out if the feed-and-expansion cistern is the source of the problem by running the programmer through its functions and checking for any

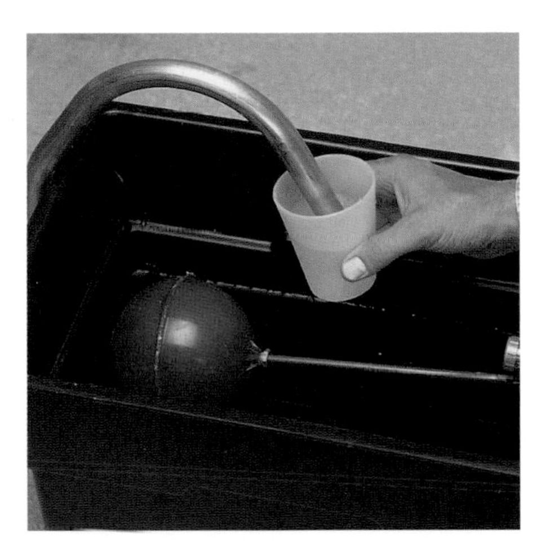

swirling movement of water in the cistern. To find out whether the vent pipe is sucking in air, submerge the end of the pipe in a cup of water. If the pipe draws up water from the cup, then air is entering the system through the pipe and causing corrosion. You will need to call a central heating engineer to rectify the problem.

Adding a corrosion inhibitor

Corrosion inhibitors are available in liquid form. In an open-vented system the liquid is added to the system through the feed-and-expansion cistern in the loft.

In a sealed system, you can inject the corrosion inhibitor into a radiator through the air vent.

If there is sludge in the system (see page 114), add a system cleaner and flush the system out before adding inhibitor.

Replacing a hot water cylinder

Hot water cylinders are mostly trouble-free, but do sometimes develop leaks or become so clogged with limescale in hard water areas that they have to be replaced.

If a hot water cylinder springs a leak, it will go on leaking as it is being fed from the main cistern. Leaks could destroy the floor (and, maybe, the ceiling) below.

Tools *Pipe grips; open-ended spanners; screwdrivers; hose; pipe clips; immersion heater spanner.*

Materials *New pre-lagged copper cylinder; PTFE tape; immersion heater fibre washer.*

Before you start Older cylinders were often lagged with a jacket, which you will need to remove. New cylinders usually come prelagged. You may have to alter the plumbing if the existing connections do not line up with those on the new cylinder.

1 Turn off the power to the immersion heater, remove the round top cover and disconnect the flex from the terminals.

2 Shut down the boiler and shut off the cold water feed to the cylinder. This pipe enters the cylinder at the bottom. If there is no gatevalve, tie the ballvalve up in the cold water tank to stop the cylinder from refilling.

3 Open the hot and cold bath taps to drain the supply pipes, but note that this does not drain the water from the cylinder.

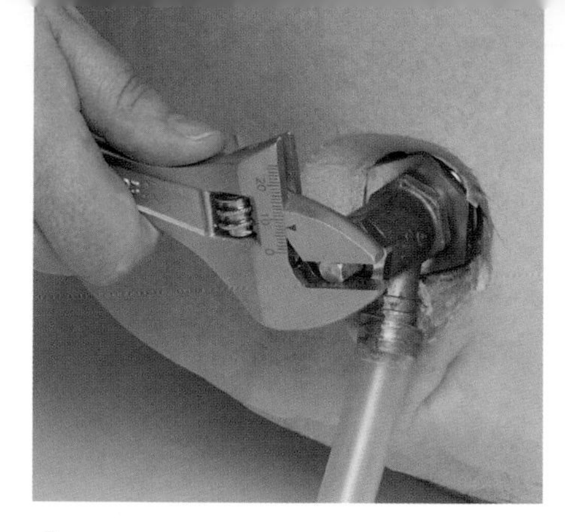

4 Attach a length of hose to the drain valve on the bottom of the cylinder. Put the other end into a drain, open the small square nut on the drain valve two turns and let the water drain from the cylinder.

5 Remove the immersion heater from the top of the cylinder by unscrewing it with a special immersion heater spanner then withdrawing it (pages 96-97).

6 Disconnect the pipes from the cylinder. Use two spanners: one to hold the securing nut on the cylinder and the other to undo the outer union nut. If you have a round-ribbed style nut use a pipe wrench instead. You will need to disconnect the cold water inlet, the hot water outlet at the top of the cylinder and the connections to and from the boiler if the water in the cylinder is indirectly heated (see drawing, right).

7 Lift out the old cylinder, being careful not to damage the ends of the disconnected pipework.

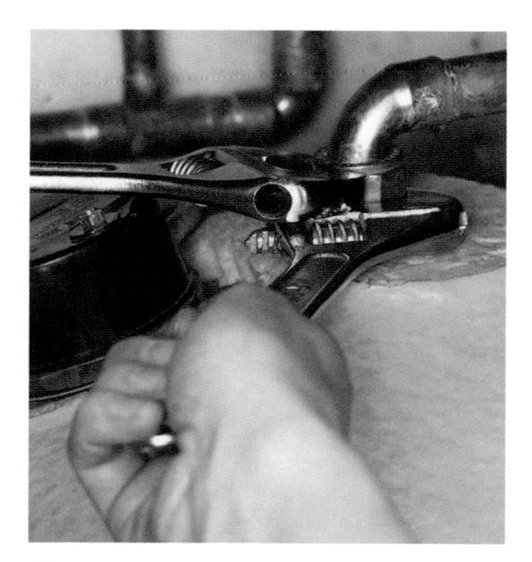

8 Wrap three layers of PTFE tape around the connection spigots on the new cylinder. Lift it back into place and reconnect the pipes, using two spanners to tighten the joint as you did when undoing the old connection.

Connections to the hot water cylinder

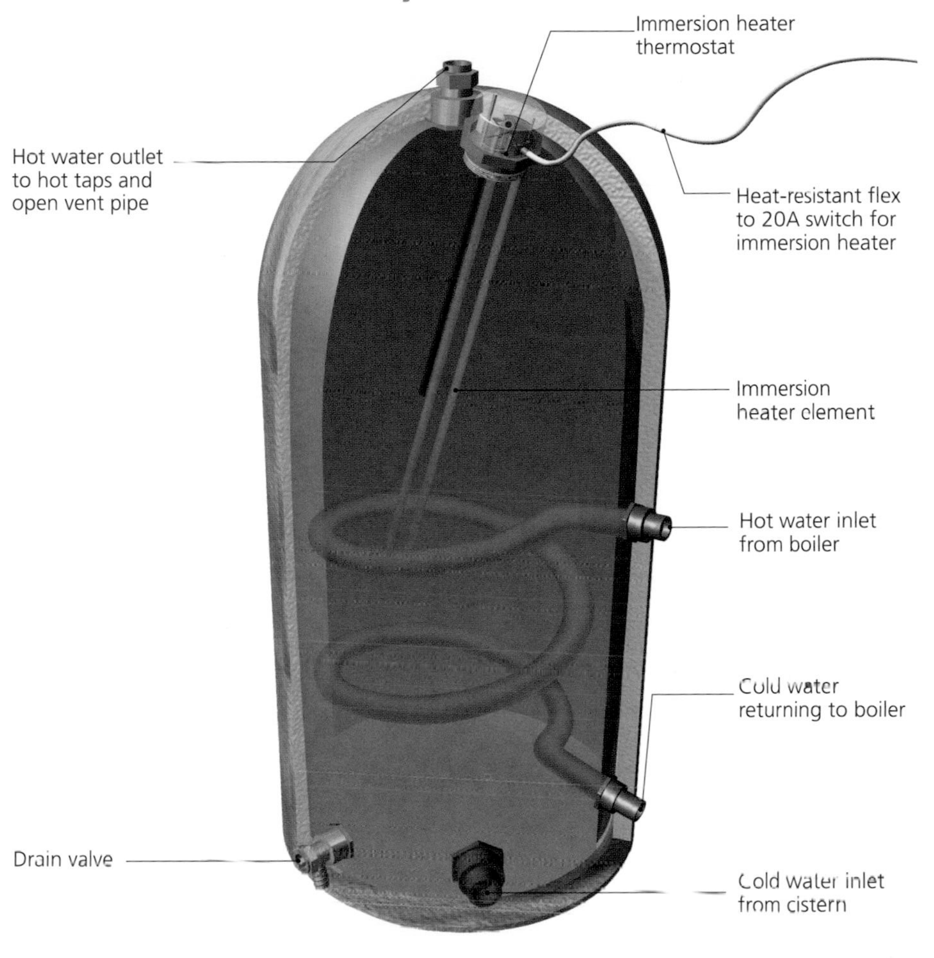

9 Refit the immersion heater (see page 98). If it was fitted with a fibre washer then replace the washer or wrap PTFE tape around the threads before you refit it. Tighten with the immersion heater spanner but do not over tighten, cylinders are thin and can crease easily.

10 Reconnect the flex to the immersion heater terminals and install the cover (see right).

11 Close the drain valve, then turn on the water supply to the cylinder. Close bath taps and check for leaks as the cylinder fills. If all is well, relight the boiler.

Having central heating installed or replaced

If you are planning to have central heating installed, read as much about heating as you can so that you can discuss it with the heating contractors.

Boilers, heat emitters and controls

Read pages 122-124 first, and then gather further information on boilers and heat emitters. You can get additional information from the advisory bodies and trade associations listed opposite. It is also worth paying a visit to your local plumbers' merchant and picking up brochures on the latest boilers and heat emitters.

Many of the larger DIY stores now sell a full range of central heating equipment – and you can find product details widely available on the Internet.

Getting quotations

Find three Gas Safe-registered contractors in your area and ask them to quote. Give them all the same outline brief, including where you would like radiators positioned, and what temperatures you wish to achieve in the rooms. A living room temperature of 21°C when the temperature outside is -1°C is normal. If you need a margin built in for extra cold weather you should say so. Make a list of any other requirements you feel are important.

1 Be wary of paying a deposit. The first payment should be when materials are delivered. Retain a small amount of the balance (2 per cent) for faults that need fixing after completion.

2 Ask the contractor to give start and completion dates.

3 Before the job starts, agree where pipes are to run and in what order they will be laid so that you can clear the room. If you want pipes to be concealed, state this before the work starts. It will cost more than surface-mounting, but is well worth the expense.

4 If several rooms will be affected, ask the contractor to finish in one room before starting in the next one.

5 Your home should be left clean and tidy at the end of each day and should be respected – for example, there should be no loud music or smoking by any of the contractors.

6 Work should comply with statutory requirements such as water supply regulations, the Building Regulations and all relevant codes of practice. Materials must meet the requirements of CEN (European) or British Standards where applicable.

7 'Making good' means filling in holes and replacing panels. Floorboards should be screwed back down to prevent creaking. Damaged boards should be replaced. Normally, however, making good does not include decoration.

8 Establish what other contractors will be required to help to complete the work – electricians, for example.

Where to install radiators

You want to get the maximum heat from the minimum number of radiators, so siting them correctly is essential. Consider the following points:

1 Radiators should be fitted in the coldest part of the room, preferably under the windows if they are single-glazed. The heat rising from the radiator will counteract the cold air falling from the glass. This produces a flow of air across the room (see right).

2 Radiators placed on inside walls opposite a window can accentuate the flow of cold air down a window and can produce a cool draught across the floor.

3 Make sure that there is at least 100mm of space between the bottom of a radiator and the floor to allow a good circulation of air and so that the floor can be cleaned.

4 At least 40mm should be left between the wall – or the skirting board – and the back of the radiator to allow air to circulate.

5 A shelf should not be placed any closer than 50mm to the top of a radiator for the same reason.

6 A radiator installed inside a decorative casing can lose a quarter or more of its output unless the casing permits a full flow of air over all of the radiator's surfaces.

Positioning a radiator To ensure the circulation of warm air around a room, it is best to install radiators under windows.

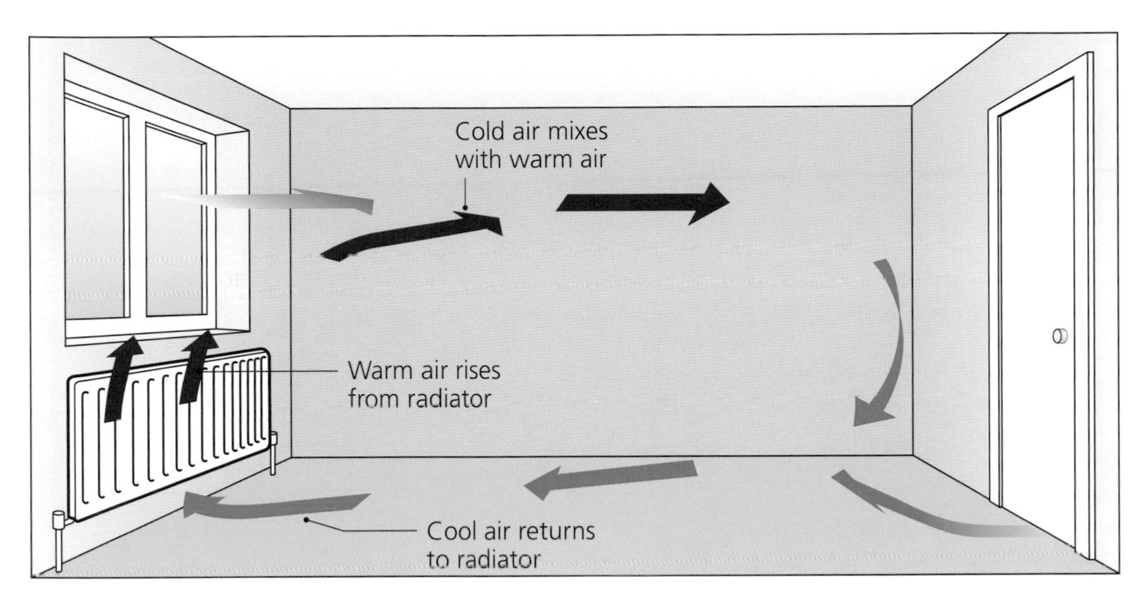

When the work is finished

The contractor must flush out the new central heating system in order to remove debris which could corrode and clog it in the future. He should then run the system to full heat and check that all the radiators heat up. The contractor must leave you all instructions and technical leaflets, and fill out guarantee cards.

Trade associations and industry bodies

Association of Plumbing and Heating Contractors (APHC)
12 The Pavilions, Cranmore Drive,
Solihull B90 4SB
tel: 0121 711 5030
www.aphc.co.uk
email: info@aphc.co.uk

Gas Safe Register
PO Box 6804, Basingstoke RG24 4NB
0800 408 5500 (freephone)
www.gassaferegister.co.uk
email: enquiries@gassaferegister.co.uk

Building & Engineering Services Association (B&ES)
(formerly Heating and Ventilating Contractors Association – HVCA)
ESCA House, 34 Palace Court,
London W2 4JG
tel: 020 7313 4900
www.b-es.org
email: contact@b-es.org

Chartered Institute of Plumbing and Heating Engineering (CIPHE)
64 Station Lane, Hornchurch,
Essex RM12 6NB
tel: 01708 472791
www.ciphe.org.uk
email: info@ciphe.org.uk

Electricians

Electrical Contractors' Association (ECA)
ESCA House
34 Palace Court
London W2 4HY
tel: 020 7313 4800
www.eca.co.uk
email: via website

National Inspection Council for Electrical Installation Contracting (NICEIC)
Warwick House
Houghton Hall Park
Houghton Regis
Dunstable LU5 5ZX
tel: 0870 013 0382
www.niceic.com
email: enquiries@niceic.com or via website

Choosing a boiler

Choosing an appropriate boiler for your household needs is crucial. Consider whether you need it to supply heating and hot water, the size of your household and how many bathrooms you have. Also think about whether it should be wall-hung or floor-standing, what type of flue is most suitable (see opposite) and the fuel it will use – oil, gas, solid fuel, LPG or electricity. A new or replacement boiler must now meet Building Regulation requirements, which demand a minimum efficiency (of 78 per cent for gas boilers, 80 per cent for LPG and 85 per cent for oil) and installation by a Gas Safe or OFTEC engineer. Programmers and thermostats must be installed, too.

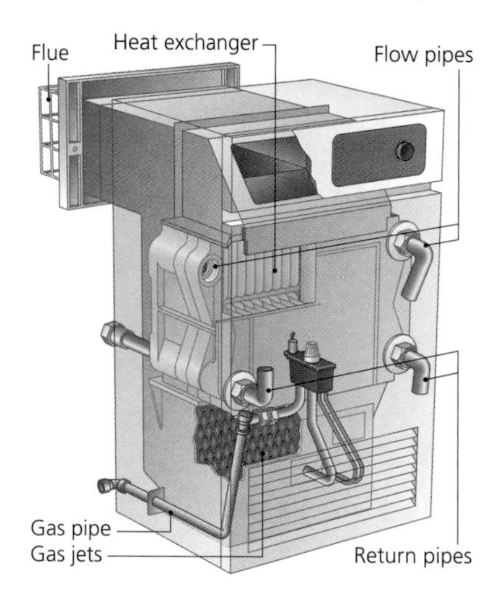

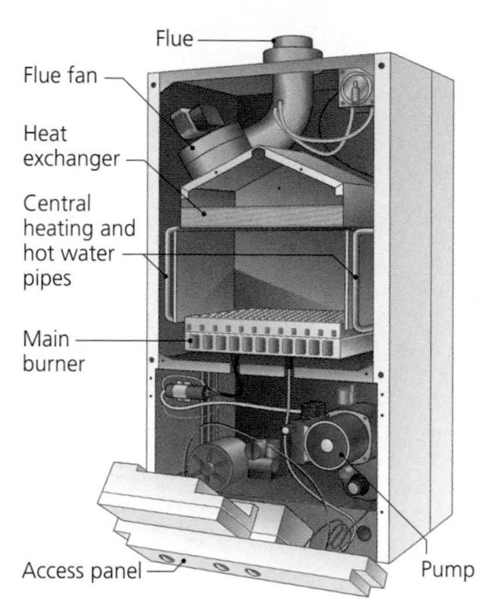

Conventional boiler The gas or oil-burner heats water in a heat exchanger, rather like a gas-ring under an old-fashioned kettle. Traditionally, heat exchangers have been made from cast-iron, but lighter aluminium and stainless steel are more commonly used now. Most modern boilers are wall-hung with balanced flues, but floor-standing models with conventional flues are still available. Most conventional boilers are designed for use on fully-pumped open-vented systems, but nearly all new boilers now have to be the condensing type – see opposite.

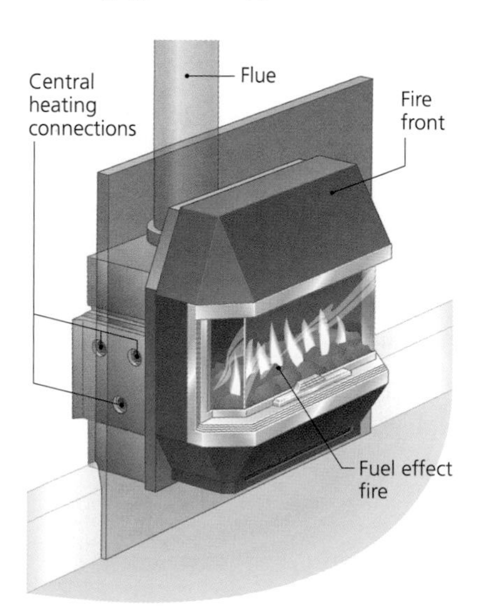

Combination boiler Also known as a 'combi' boiler, this is a central-heating boiler and multi-point water heater all in one. Hot water for the radiators is heated in its own sealed circuit but the boiler also heats cold water from the mains, delivering it on demand to the hot water taps around the house. The main advantages are the savings in space – no hot water cylinder or cisterns – a constant supply of hot water and better water pressure in showers. By altering the cold-water plumbing, there can also be drinking water at all cold taps. The disadvantages are the cost of the boiler and low flow-rates if more than one hot tap is being used. New combination boilers are nearly always the condensing type – see opposite. Most can be used with gas or LPG, though oil-fired options are also available.

Back boiler This is a heat exchanger located behind a gas fire. Although many still exist in older houses, they are rarely fitted new. They work in much the same way as a conventional boiler, sending hot water to radiators on a central heating circuit and to a hot water cylinder, but need a conventional open flue, suitably lined for the fuel being used. A back boiler can be used with a fully pumped system, and condensing versions are available. The firefront may be inset into a fireplace or may protrude into the room.

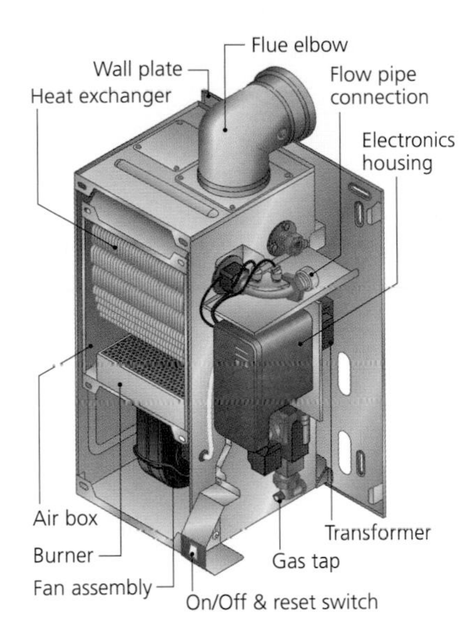

Condensing boiler With a larger heat-exchanger than a conventional boiler, a condensing boiler is designed so that the water returning from the heating system is used to cool the flue gases, extracting extra heat that is normally lost through the flue. Often known as 'high efficiency' boilers, they are meant to be used with a fan-assisted balanced flue and in a fully-pumped system. When the flue gases are cooled, water vapour will condense and so a pipe has to be installed to drain this water away.

Condensing boilers work best with lower system water temperatures, but even with normal radiator temperatures, the efficiency will be significantly greater than with a conventional system. All new boilers now have to be the condensing type in order to meet Building Regulations – unless it is physically impossible to fit one (no possibility for a condensate drain, for example). Condensing boilers are available for use with either gas, LPG or oil. Combination condensing boilers are available, too.

TWO TYPES OF FLUE

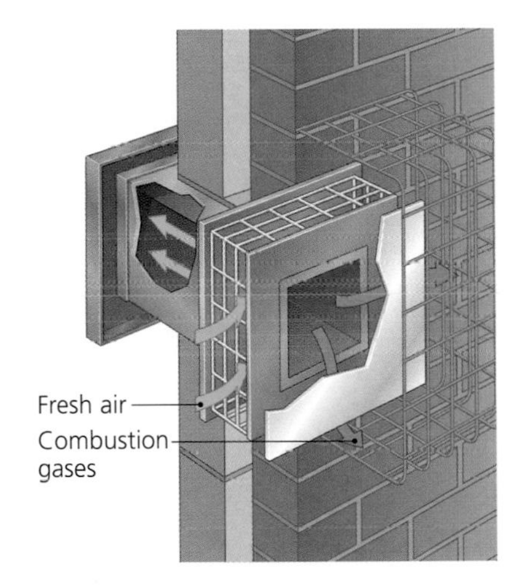

Balanced flue In order to work properly, this two-part duct allows the combustion gases to escape and fresh air to enter. The flue is sealed so that no combustion gases can enter the room where it is installed – its other name is a 'room-sealed flue'. With a natural-draught balanced flue (as above) the boiler must be installed on an outside wall, so that the flue passes directly through the wall. With a fan-assisted balanced flue, the boiler (which contains an electric fan) can be mounted on any wall, and is connected by a duct to a flue that can be on an outside wall or pass through the roof. Fan-assisted balanced flues are more efficient, but noisier, than natural-draught flues.

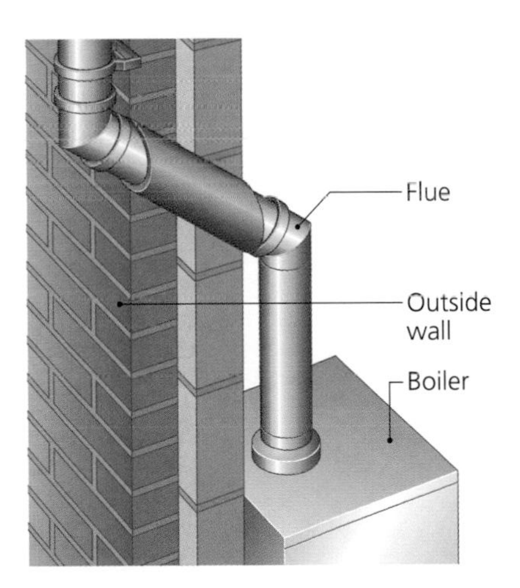

Open flue This can be either a lined existing chimney or a new circular duct installed in an outside wall. The flue will take only the combustion gases, so the fresh air supply for the burner must come from the room. Consequently, special ventilators or grilles will need to be installed on outside walls.

Choosing radiators and other heat emitters

Though most people's first choice is a radiator, there are many other heat emitters that can be connected to central-heating pipes.

These include fan convectors, trench-duct heaters, skirting heaters and under-floor heaters. When mixing different types of heat emitter on the same system, fit thermostatic valves to each in order to allow full, individual control.

Radiators Despite the name, only a tiny proportion of heat given off by a radiator is emitted from the front through radiation. If you put your hand just a few inches from the front the heat is negligible. Most of the heat is given out from the top by convection.

To work properly a radiator must have a good flow of air passing from the bottom on the front and back surfaces. There must be at least 100mm clearance from the floor for air to enter and 40mm at the top.

Old style plain panel radiators have now been almost completely superseded by convector radiators, which have metal boxed fins welded to the hidden faces of the panels. They act as chimneys for hot air, almost doubling the heat output and making it possible to fit smaller radiators. There are many different styles of tubular radiator available, from modern interpretations of a traditional Victorian style (above left) to quirky wall-mounted spirals (above right).

Underfloor heaters Burying pipes under concrete floors has gained in popularity. Plastic pipe is laid in a continuous loop and carries hot water under the floor. The pipes must be fitted on top of under-floor insulation and are normally covered with a sand-and-cement screed which helps to spread the heat evenly. This is an ideal system for use with a condensing boiler, because it works well at low temperatures.

Trench-duct heaters If windows go down to the floor, trench heaters can be installed. A pipe fitted with fins runs along one side of a trench in the floor. A dividing plate along the centre of the trench separates the hot air rising from the pipe from the cooler air returning to be reheated.

Skirting heaters

Small metal convectors run round the room just above or in place of a skirting board. This system is good for background heating and it gives an even spread of heat, which can help to prevent condensation on walls. However, it is not usually powerful enough to heat a room in very cold weather.

Fan convectors Use a fan convector where there is not enough wall space for a radiator. Special kick-space models are made to go under kitchen base units (below). Low voltage versions are available for bathrooms. Air curtain models can be installed above doors and wall units, and some models sink into the floor.

An electrical fan blows air across copper fins, which are heated by hot water from the central-heating circuit. A filter in the air intake traps dirt. This should be regularly cleaned to ensure maximum performance and to prevent noise.

A

B

C

D

E

F

G

H

I

L

M

O

P

R

S

T

U

V

W

Acknowledgments

All images in this book are copyright of the Reader's Digest Association, Inc., with the exception of those in the following list.

The position of photographs and illustrations on each page is indicated by letters after the page number:
T = Top; **B** = Bottom; **L** = Left; **R** = Right; **C** = Centre
Front cover Shutterstock/Milos Luzanin; istockphoto/Mark Stay; **19 CR** Worcester, Bosch Group; **64 BL** 'Tratto' by Ideal Standard; **CR** www.bathstore.com/'XT' wash basin mixer; **TL** 'Cliveden' pillar taps by Armitage Shanks; **65 TL** 'Hathaway' by Armitage Shanks; **TR** 'Academy' by Ideal Standard; **BL** 'Palladian' by Ideal Standard; **BR** 'Millenia QT' by Armitage Shanks; **66 T** 'Academy & Kyomi' by Ideal Standard; **67 TL** 'Plaza' by Ideal Standard; **TR** 'Kyomi' by Ideal Standard; **CL** www.bathstore.com/'Square' designer basin; **CR** 'Meadow' handrinse basin by Ideal Standard; **B** 'The Space Studio Bidet' by Ideal Standard; **74 B** www.bathstore.com/'Delta Corner'; **75 B** 'Plaza' close coupled WC by Ideal Standard; **85 B** 'Calista' Trevi showers by Ideal Standard; **88 TL, BL** GE Fabbri Limited; **112** GE Fabbri Limited; **113 TL, BL** GE Fabbri Limited; **124 TL, TC, TR** www.bisque.co.uk; **BR** Myson Radiators

Reader's Digest DIY Plumbing and Heating
This edition published in 2012 in the United Kingdom by Vivat Direct Limited (t/a Reader's Digest), 157 Edgware Road, London W2 2HR

First published as **Reader's Digest Plumbing and Heating Manual** in 2005

Project Editor Jo Bourne

Art Editor Sailesh Patel

Consultant David Holloway

Editorial Director Julian Browne

Art Director Anne-Marie Bulat

Managing Editor Nina Hathway

Trade Books Editor Penny Craig

Picture Resource Manager
Sarah Stewart-Richardson

Pre-press Account Manager Dean Russell

Production Controller Jan Bucil

While the creators of this work have made every effort to ensure safety and accuracy, the publishers cannot be held liable for injuries suffered or losses incurred as a result of following the instructions contained within this book. Readers should study the information carefully and make sure they understand it before undertaking any work. Always observe any warnings. Readers are also recommended to consult qualified professionals for advice.

Typesetting, illustration and photographic origination
Hardlines Limited, 17 Fenlock Court, Blenheim Office Park, Long Hanborough, Oxford OX29 8LN

Origination FMG
Printed and bound in China

Reader's Digest DIY Plumbing and Heating is based on material in **Reader's Digest DIY Manual** and **How Everything in the Home Works**, both published by The Reader's Digest Association, Inc.

We are committed both the quality of our products and the service we provide to our customers.
We value your comments, so please do contact us on **0871 351 1000**, or via our website at **www.readersdigest.co.uk**

If you have any comments about the content of our books, email us at **gbeditorial@readersdigest.co.uk**

ISBN 978 1 78020 128 3
BOOK CODE 400-598 UP0000-1